John Ball

North Tyrol, Bavarian, and Salzburg Alps

John Ball

North Tyrol, Bavarian, and Salzburg Alps

ISBN/EAN: 9783743435438

Printed in Europe, USA, Canada, Australia, Japan

Cover: Foto ©Andreas Hilbeck / pixelio.de

Manufactured and distributed by brebook publishing software (www.brebook.com)

John Ball

North Tyrol, Bavarian, and Salzburg Alps

Ball's Alpine Guides

NORTH TYROL, BAVARIAN

AND

SALZBURG ALPS

BY

JOHN BALL, F.R.S., M.R.I.A., F.L.S., &c

LATE PRESIDENT OF THE ALPINE CLUB

LONDON
LONGMANS, GREEN, AND CO.
1873

LIST OF MAPS.

Key Map   *To be pasted inside the cover at the beginning*

The Eastern Alps—General Map   .   .   .   *To face title-page*

The Berchtesgaden District   .   .   *To face page* 71

ABBREVIATIONS AND EXPLANATIONS.

The following are the chief abbreviations used in this work :—

hrs., m.—for hours and minutes. When used as a measure of distance, one hour is meant to indicate the distance which a tolerably good walker will traverse in an hour, clear of halts, and having regard to the difficulty of the ground. In cases where there is a considerable difference of height, the measure given is intended as a mean between the time employed in ascending and descending, being greater in the one case and less in the other.

ft., yds.—for feet and yards. The heights of mountains, &c., are given in English feet above the level of the sea, and are generally indicated in the manner usual in scientific books, by the figures being enclosed in brackets, with a short stroke.

m.—for mile. Unless otherwise expressed, distances are given in English statute miles.

rt., l.—for right and left. The right side of a valley, stream, or glacier, is that lying on the right hand of a person following the downward course of the stream.

The points of the compass are indicated in the usual way.

Names of places are referred in the Index to the pages where some useful information respecting them is to be found.

Throughout this work the reader is frequently referred for further information to the Section and Route where this is to be found. When the reference is made to a passage occurring in the same Section, the Route alone is mentioned.

ALPINE GUIDE.

CHAPTER XII.

SUABIAN ALPS.

SECTION 41.	Route D — Munich to Innsbruck, by Partenkirch . . . 26
ALGAU DISTRICT.	Route E — Munich to Partenkirch, by Ammergau 29
Route A — Immenstadt to Landeck, by the Algau and Schrofen Pass 3	Route F — Partenkirch to Imst in the Innthal 31
Route B — Bregenz to Oberstdorf in Algau, by the Bregenzer Wald 5	Route G — Munich to Innsbruck, by the Walchen See . . . 34
Route C — Bregenz to Sonthofen, by the Bregenzer Wald . . . 8	SECTION 43.
Route D — Sonthofen to Reutte in the Lech Thal 9	KREUTH DISTRICT.
Route E — Feldkirch to Reutte, by the Walserthal and Lech Thal . 11	Route A — Munich to Innsbruck, by Wildbad Kreuth 35
	Route B — Munich to Innsbruck, by Kufstein and Schwaz . . 40
SECTION 42.	Route C — Munich to Brixlegg in the Innthal, by Schliersee . . 43
ZUGSPITZ DISTRICT.	Route D — Munich to Kufstein, by Baierisch-Zell 45
Route A — Landeck to Innsbruck, by the Innthal 14	Route E — Tegernsee to Mittenwald, or Partenkirch . . . 46
Route B — Augsburg, or Lindau, to Innsbruck, by Füssen and Lermoos 20	Route F — Walchensee to Pertisau on the Achensee 49
Route C — Reutte to Landeck, by the Lech Thal 24	Route G — Scharnitz to Jenbach in the Innthal 50

To THE eye of the geologist, the general arrangement of the Eastern Alps is more symmetrical than that of the portions of the great chain described in the preceding volumes of this work. A central mass, composed almost exclusively of crystalline rock, extends nearly due E. and W. from the frontier of Switzerland to that of Hungary, flanked to the N. and the S. by ranges of sedimentary rocks, amongst which jurassic limestones predominate, especially on the southern side of the main chain. The principal valleys, and more particularly those immediately connected with the central chain, are parallel to it in their general direction, whereas those of the exterior ranges are less regularly disposed, though even there the relation between the minor ridges and valleys and the adjacent portions of the main chain is closer than can reasonably be attributed to accidental causes.

In the present chapter is described

M. T. B

the mountain region lying north and west of the Inn, extending westward to the Lake of Constance, and northward to the plain of Bavaria. The larger part of this region belongs to Bavaria, and hence the entire group is sometimes known as the Bavarian Alps. That designation being obnoxious to the people of the Austrian province of Tyrol and Vorarlberg when applied to the portions which they inhabit, it seems advisable to employ the somewhat more vague term, *Suabian Alps*, in default of any more generally recognised geographical denomination. The country described in this chapter abounds in attractions for the artist and the sportsman, but, save a few spots hereafter noticed, it is little visited by English travellers. Except in the neighbourhood of Bregenz the mean temperature is lower than in any other part of the Alps, and hence it happens that although but few of these summits rise above 9,000′, and the general level of the ridges is much below that limit, snow lies throughout the year in considerable masses, and small glaciers are found on several of the higher mountains.

SECTION 41.

ALGAU DISTRICT.

IN viewing the chain of the Alps from the Bavarian plain the first considerable valley on the west side is that of the Iller, the upper portion of which, belonging chiefly to Bavaria, is known as the Algau. It is enclosed by mountains which usually approach to, but seldom surpass, the limit of 8,000 ft. On the east and south sides, where they divide the basin of the Iller from that of the Lech, the summits assume the bold and picturesque forms characteristic of the dolomite and jurassic limestone; while to the N. and W. the prevailing rocks are of cretaceous or tertiary age, and constitute a considerable highland region known as the Bregenzer Wald, extending to the Lake of Constance, and drained in great part by streams that flow through that reservoir into the Rhine. The frontier between the Algau and the Austrian province of Vorarlberg keeps pretty closely to the watershed between the Danube and the Rhine, except at the SW. extremity of the first-named district, where the upper part of the Breitach glen belongs to Vorarlberg. The latter province is divided into two nearly equal portions by the high road from Feldkirch to Landeck, which along with the southern portion of the province was described in § 34. Eastward of the Iller the next considerable stream that descends from the Suabian Alps into the plain of Bavaria is the Lech. From Füssen, where it issues from the mountains, it follows a tolerably direct northerly course to Augsburg; but the upper course of the same river circles round to WSW. and divides the mountain range at the head of the Algau from the parallel range which forms the N. boundary of the Inn valley between Imst and Landeck, and extends westward from the latter place to Bludenz.

A glance at the geological map which accompanies the second part of this work shows that here, as well as elsewhere in the Alps, the principal ridges and the corresponding depressions are approximately parallel to the lines marking the outcrop of the strata. Whatever may have been the causes that determined the direction from WSW. to ENE. of the great ridges and valleys throughout Switzerland, they have evidently extended to the district here described, while there are no less manifest indications of a line of strike running nearly due E. and W. which characterises a large portion of the Eastern Alps. The valley of the Lech forms the natural boundary of this district to the E. and SE., and to the SW. it is natural to include in it the Walserthal and the mountains between that and the main branch of the Ill leading through the Klosterthal.

Thus defined the present district is bounded by the road from Bregenz to Feldkirch in the valley of the Rhine, and that which thence follows the Ill to Stuben at the foot of the Arlberg Pass. Then traversing the low pass leading to the upper valley of the Lech it follows that stream to Füssen; the northern boundary being marked by the plain of Bavaria. The road from Kempten to Füssen and Reutte, though included in the above limits, is more conveniently described in the next section.

Good quarters for pedestrians are found in most of the chief villages of this district, besides which there are many small establishments connected with mineral springs, where the mountaineer may resort during the summer months. Four or five days may be very well employed in the picturesque glens of this district, and may serve as a preparation for more arduous excursions in the higher region of the Tyrol Alps. The best head-quarters for the mountaineer are found at Oberstdorf in the Algau, and at Schoppernau in the Bregenzer Wald.

Route A.

IMMENSTADT TO LANDECK, BY THE ALGAU AND SCHROFEN PASS.

	Stunden	Eng. miles.
Sonthofen	2¼	6¾
Oberstdorf	3	9
Lech	7	17¼
Stuben	3	9
Landeck	8	24
	23¼	66¼

The most convenient point for approaching the valley of the Iller (Algau) or that of the Lech (Rte. D) is

Immenstadt (Inn : Post, kept by Hilsenbeck, very fair), a small neat town on the rly. from Augsburg to Lindau, standing (about 2,080′) at the point where the valley of the Iller opens into the plateau of Bavaria. The pretty *Alp See*, lying W. of the town on the S. side of the rly. to Lindau, and several ruined castles, contribute to adorn the neighbouring scenery. A favourite excursion is the ascent of the *Grünten* (5,558′), a few miles E. of the town. The summit, reached in about 3½ hrs., commands an extensive panorama, including the Cathedral of Ulm to the N., the Lake of Constance, and a wide circuit of Alpine peaks to the S. and E. There is an inn with thirty beds about 20 min. below the summit. The eocene rocks at some points abound in fossils, and ores of iron are worked near the top of the mountain. The traveller may descend in 2 hrs. direct to Sonthofen. A guide is scarcely required.

After passing *Seyfriedsberg* the road crosses the Iller, and soon after reaches *Sonthofen* (Inns : Engel ; Hirsch ; Adler), a small market town (2,452′), the principal place of the Algau. Omnibuses run twice a day between this and the rly. station at Immenstadt. Here the post road leading to Tyrol ascends eastward along the valley of the Ostrach (Rte. D), while the country road through the main valley keeps due S. on the E. side of the Iller, but at some distance from its banks. After passing Altstetten, and other small villages, as well as two ruined castles, the traveller reaches the point where the stream of the Iller is formed by the junction of the torrents *Trettach, Stillach*, and *Breitach*, issuing from the three principal valleys of the Algau Alps. A short distance beyond the junction stands

Oberstdorf (2,594′), on the tongue of land dividing the Trettach from the Stillach. This neat village, frequented in summer by visitors from Augsburg and Nuremburg, was in great part burned down in 1865. It affords the best head-quarters for mountain excursions in Algau, with fair accommodation at two inns (Sonne ; Mohr), and a bold and efficient guide in Blattner; a gamekeeper (Jagdbehülfe) named Franz Schafhittl, is less to be recommended for difficult ascents.

Of the numerous excursions that may be made from Oberstdorf, the following deserve special mention.

The wild stony glen of the Trettach leads nearly due S. from the village to the *Christlessee*, a basin of dark-blue water at the S.E. base of the *Mädelegabel* (8,674′), the highest of the Algau Alps. The extent of snow is far greater than would be expected at so moderate an elevation. The *Mädelejoch* Pass, over the ridge E. of the Mädelegabel, leads in 7 hrs. from Oberstdorf to Holzgau, on the Lech (Rte. E), from whence Landeck may be reached by the Kaiserjoch in rather less time, but by a less interesting route, than that described below.

The lateral glen of the *Oythal*, joining the Trettachthal near Oberstdorf, contains much interesting scenery. The lower portion is a mere defile, and at the narrowest part the torrent, springing over two successive shelves of rock, forms a remarkable double waterfall. From the upper end of the glen a track leads over a pass on the N. side of the Rauheck (7,824′) to Hornbach, in a lateral valley of the Lechthal. Another track which bears to the l., somewhat NE., leads to the head of the Hintersteiner Thal, and from the ridge affords the easiest access to the summit of the Hoch Vogel (8,487′), further noticed in Rte. D.

An active pedestrian may in a single day make an agreeable round of very varied scenery by following a track that mounts due E. from Oberstdorf; he should then descend by the Wengen Alp into the Hinterstein glen, follow the Ostrach to Hindelang, and return by Sonthofen.

The valley of Breitach, which may be visited from Oberstdorf, is described in Rte. B.

The way to the Schrofen Pass lies through the valley of the Stillach, probably the most interesting of the Alpine glens of Algau. The bold pyramidal peak rising S. of the village between the Trettach and Stillach, called *Schrofen*, is not to be confounded with the pass of the same name. The car-road seems to be practicable as far as Einödsbach. On a terrace of the mountain to the rt. lies a solitary tarn called Freyberger See. After lying for some way through a defile, the track reaches the hamlet of Faistenau. To the W. rises the *Schlappolt* (6,405′), sometimes ascended for the sake of the extensive view which it commands, but the mountaineer will prefer some one of the higher dolomite peaks forming the frontier line of Tyrol, especially if he has not had previous experience of the somewhat peculiar pleasures of dolomite climbing.

A little open plain with a few scattered houses, called Birgsau (3,185′), is succeeded by a narrow defile, which leads to *Einödsbach*, the highest hamlet in the valley. Here a ravine mounts southward to the Schneeloch, a wild hollow between the Mädelegabel and the *Biberkopf* (8,543′). An account of the somewhat difficult ascent of the latter peak, effected in 1857 by Dr. Holler, with the above-named guide, Blattner, is given in the second volume of the Proceedings of the Austrian Alpine Club.

Above Einödsbach the glen of the Stillach mounts towards SW., and is thenceforward known as the *Rappenalpenthal*. To the rt. rises the Gaishorn, and to the l., at the head of the glen, the *Gross Rappenkopf* (8,226′) and *Klein Rappenkopf* (7,471′). In approaching the dolomitic peaks that form the S. boundary of Algau, the geologist will not fail to remark the outcrop of a zone of friable slate which everywhere underlies the dolomite, and is conspicuous, even at a distance, by its more abundant vegetation. This, sometimes called Algauschiefer, was formerly referred to the lias, but is now usually considered to belong to the trias.

The path makes a sharp zigzag to the l. before the final ascent to the *Schrofen Pass* (5,569′)—4 hrs. from Oberstdorf, the lowest of those connecting the Algau with the upper valley of the Lech.

The shortest way to Landeck is to take the bridle-path bearing to the l. in descending from the pass. This leads down the valley of the Lech in 2½ hrs. to *Steg* or *Stögen* (3,725′), with two tolerable inns. By a lateral glen opening S. the traveller may thence reach in 1½ hr. the village of *Kaisers* (poor inn), and in 5 hrs. more traverse the *Kaiserjoch* to *Pstneu*, a village on the high road of the Arlberg, 4¾ hrs. from Landeck. The way lies at first nearly due S. from Kaisers, but where the head of the valley bends to ESE. the path keeps to the rt. At the summit it is marked by poles. The rte. described below is easier, more attractive, and but little longer.

[From Lechleiten the traveller bound for the Bregenzer Wald may reach *Krumbach* (Rte. E), the highest village in Vorarlberg, 5,481 ft. above the sea, which is separated by scarcely any perceptible ridge from the head of the Bregenzer Ach. But this would be a circuitous course for a traveller going from Oberstdorf, the more direct way being that by the Haldenwanger Eck. mentioned in Rte. B.]

The way from the Schrofen Pass to Stuben bears to the rt. during the descent, then crosses the stream to Wart, and ascends along the rt. bank to *Lech*, (4.094′). a hamlet (with a poor inn) standing at the opening of the Zürserthal. The uppermost end of the Lech Thal, extending several miles into the Vorarlberg, and locally called *Tannberg*, is enclosed by the highest summits of this district, the chiefs of them being the *Rothewand* (8,842′), to the N., and the *Schafberg* (8,774′) to the S., both supporting considerable glaciers. To the southward, through the opening of the *Zürserthal*, the snowy summit of the *Roglerspitze* (7,660′) shows its double point. An easy pass connects the short glen of the Zürserthal with the upper end of the Klosterthal, and by that way

Stuben (Post, fair, not cheap), on the W. side of the Arlberg Pass, may be easily reached in 3 hrs. from Lech by a low and easy pass at the head of the Zürserthal. If bound for Landeck, avoid Stuben, and take a well-marked path to the l. after crossing a bridge

over the main stream. In 1 hr. from the summit the old Arlberg road is reached, ¼ hr. more takes you to the new road, and 20 m. additional to the summit of the Vorarlberg Pass.

The high road to Landeck (Inns: Schwarzer Adler; Goldener Adler; Post) is described in § 34, Rte. A.

ROUTE B.

BREGENZ TO OBERSTDORF IN ALGAU,
BY THE BREGENZER WALD.

	Stunden	Eng. miles
Alberschwendy	3½	10¼
Schwarzenberg	3	9
Mellau	2½	7½
Schoppernau	3	9
Mittelberg	4	10
Oberstdorf	4	11
	20	57

The traveller who has reached the Lake of Constance by any one of the numerous railways that converge upon its shores, may commence a tour in the German Alps in a very agreeable manner, by following the route here described, passing on the way through some of the most pleasing portions of the two main valleys included in the present section.

Bregenz (Inns: Œsterreichischer Hof, good and clean; Goldener Adler, or Post, good; Schwarzer Adler; Krone) is placed in frequent and rapid communication with the principal places on the Lake of Constance by the lake steamers, and besides this it will soon be connected with Coire and Lindau by railway. Families intending to make a carriage tour in Tyrol may find it a good plan to engage a vehicle here. A Lohnkutscher, named Kiefer, is highly recommended. 'He has good carriages and horses, and knows the Tyrol well; he and his man are very careful drivers.' [G.H.S.]

Bregenz is a place of great antiquity (Brigantia of the Romans), and the upper town stands on the site of the Roman *castrum*, one of whose gateways is still preserved. Several objects of antiquity found in the neighbourhood are still preserved in the Vorarlberger Museum. The position is very agreeable, and the little town has more wealth and resources than might be supposed from its small size and population. These arise chiefly from the timber trade, including poles for vines, of which more than 2,000,000 are annually exported to the wine districts surrounding the lake. The wooded hill rising E. of the town is called *Pfannenberg* (3,485'), and also *Gebhardsberg*, from an ancient chapel standing on the S. end of the ridge. A German traveller advises strangers approaching the town by the Feldkirch road to ascend the hill before entering the town, carefully abstaining from turning round until they have reached the Gebhardskirche, and so take their first view of the wide expanse of the lake from the open window of a portion of the ruined castle, now used as an inn. The botanist may find the rare *Carex Gaudiniana* in a marshy hollow near the pathway called Siechelsteige, and *Dianthus cæsius* on the Klausberg.

It has been said in the introduction that the extensive mountain region extending S. and E. of Bregenz, between the Rhine and the Algau, is called *Bregenzer Wald*. Under that name are included districts whose diversity, obvious to the ordinary observer, closely corresponds with differences of geological structure. There is however one common characteristic arising from the disposition of the mountains in nearly parallel ridges running from W. to E., or from WSW. to ENE. The minor streams follow the direction of the troughs between the parallel ridges, while the *Bregenzer Ach*, which unites to itself nearly all of them, follows a strangely sinuous course, as it alternately follows the direction of a trough, or has cut its way through the ridge which separates this from the adjoining depression. The northern portion, locally known as the Aeussere Bregenzerwald, lies within the tertiary (molasse) formation. Here the streams have cut extremely deep trenches, for the most part impassable, so that the villages and scattered houses are to a great extent isolated, and near neighbours can reach each other only by a long detour. This circumstance has doubtless contributed to maintain many local peculiarities in the customs, and even in the language, which is said to approach to the form of old Hoch Deutsch, in which the legendary poem of the Nibelunge Noth has been handed down. Farther south the rocks, belonging to the age of the English greensand, are more compact, the ridges are higher, and the valleys, even when narrow, are usually accessible to a pedestrian. The Jura limestone, which gives their peculiar forms to the higher mountains of this district, is limited to the extreme S. end of the valley, and to a small isolated group above Mellau, whose highest summit is the *Mittagspitz* (6,851').

The pedestrian, who is sure of tolerably good accommodation throughout the district, owing to the number of mineral springs resorted to for drinking or for baths, may select for himself among the very numerous paths and cart-tracks that lead from one village to another, but those who wish to avail themselves of country carriages should prefer the route here described. The Bregenzer Ach descends through a deep gorge to the level of the lake on the S. side of the Gebhardsberg, but the carriage-road at first avoids it altogether, and follows the base of the hills on the E. side of the Rhine as far as *Schwarzach*, a village (with a fair inn) standing at the point where a stream of the same name enters the valley. The road ascends in zigzags, and after

ROUTE B.—BREGENZER WALD.

passing a solitary inn at the summit level, descends into a green basin, where stands the little village of *Alberschwendy* (2,178′). Here, as well as elsewhere in the Bregenzer Wald, the population is scattered in single houses or small groups, and even in populous places the village consists of but twenty or thirty houses, gathered round the church, the parsonage, and the inn. The pedestrian bound for Schwarzenberg may save nearly 1 hr., and gain a pleasant walk, by crossing the hill called Lorena, round whose E. base the road makes a wide circuit, descending, and for the first time approaching the ravine of the Bregenzer Ach. In one place a rough road is carried down to a bridge, and then up the steep opposite slope to *Lingenau* (Rte. C), on the N. side of the Sübers Bach, one of the principal affluents of the main stream. S. of that torrent, and E. of the Bregenzer Ach, is *Egg*, a commune with two establishments of mineral baths. Keeping to the W. side of the valley, the road, bearing SSW., reaches *Schwarzenberg*, (Inns: Hirsch, good; Sächsischer Hof), the birthplace of Angelica Kaufmann, whose name is held in great honour here. An altar-piece by her is shown in the church, and other works are, or used to be, preserved in a house, formerly the inn (Zum Lamm), whose owner was a relative of the artist. The path to Dornbirn (§ 34, Rte. A) over the *Hochälple* (4,800′) commands fine views.

Beyond Schwarzenberg the valley is narrowed to a defile, forming the boundary between the Outer and Inner Bregenzer Wald. Beyond the defile a branch road crosses a bridge and ascends to *Bezau*. On the hill above the village, at a place called Bezeck, the ancient legislative assembly of the Bregenzer Wald used to assemble on a high stage, accessible only by ladders, which were removed when the assembly was complete, and replaced only after a final decision on all matters in debate had been arrived at. South of this is *Reute*, with an establishment of mineral waters, one of the best kept and most frequented in the district. The scenery becomes bolder as the traveller approaches *Mellau*, a village with a very fair country inn. 'Der Messner' is recommended as a guide.

To the SE. rises the high limestone ridge of the *Canisfluh* (6,696′), and due S., the rather more distant parallel ridge of the *Mittagspitz* (6,851′). Two fine waterfalls—the Fluhbach and Kobelbach—both deserve a visit. The latter, 300 ft. in height, is in a lateral glen called *Mellenthal*, through which the valley of the Rhine may be reached, either at Rankweil by descending through the *Laternserthal*, or near Dornbirn by following a glen whose course is nearly due N. These three glens radiate from the *Hohe Freschen* (6,571′), an isolated summit 'commanding a charming view of the whole Vorarlberg and the high Alps.' [E. M.]

From Mellau the road after crossing to the rt. bank runs due E., close under the menacing precipices of the Canisfluh, to *Schnepfau* (2,364′). Here the valley bends to SE. and S. till it widens at the village of *Au* (with a good inn, Zum Rössli), opposite the opening of the *Argenthal*, a lateral glen through which the traveller may reach Rankweil by the *Laternserthal*, or traverse the ridge separating it from the Walserthal by a pass leading to Sontag (Rte. E). At *Damils*, the chief village of the Argenthal, there is a tolerable inn. The carriage-road through the main valley comes to an end at

Schopernau (2,954′), where the little village inn (Krone) is said to be the best in the Bregenzer Wald. The traveller bound for the Algau may here choose between various mountain tracks, but the easiest and most direct is that of the *Starzljoch*, by which he will reach in 4 hrs., or less, the Baths of *Mittelberg*, at the head of the glen of the Breitach, one of the chief affluents of the Iller. Although geographically a portion of the Algau, this branch of the valley belongs to Vorarlberg, and above the frontier of Bavaria is called *Mittelberg*, and also Walserthal, because of its frequent relations with the valley of that name noticed in Rte. E. Though the population of the Mittelberg is rather considerable for so remote a district, there is nothing that can be called a village, as the houses are scattered throughout the valley. Near Hirschegg the track turns away from the stream which descends to the lower valley through an impassable cleft. After passing *Rietzlen* (2,815'), the path descends the rocky barrier through which the stream has cut a channel, and rejoins its banks near to a wayside inn where a wooden bar marks the Bavarian frontier. Below this spot, called Schanzl, the glen becomes a defile which opens out again at the junction of the Starzlach with the Breitach. Here the path to Oberstdorf bears to the rt., crossing the lower terrace of the mountain range that divides the Breitach from the Stillach, and after traversing the latter stream reaches the village, for which see Rte. A.

A longer way from Schopernau to Oberstdorf than that above pointed out, but offering grander scenery, is to follow the stream of the Bregenzer Ach. A rather steep track mounts along the rt. bank to *Hopfreben*, where there is a sulphurous spring frequented by the country people. The defile becomes still narrower beyond this point, but it opens out suddenly at

Schrecken (3,806', Mayr—probably higher?), the highest village in the valley, much exposed to avalanches, with a tolerably good new inn. The village stands at the S. base of the *Widderstein* (8,294'), a rugged range whereon the botanist finds many rare plants; *e.g. Crepis montana* and *C. hyoseridifolia*. Owing to the peculiarity in the orography of this district noticed in the introduction, the head of this valley is connected by very low passes with the head of the Walserthal to the W., and with a short branch of the Lech Thal, where stands the village of *Krumbach*, to the E. The latter pass affords the most direct and easiest route from the Bregenzer Wald to the valley of the Lech. There is another pass leading nearly due S. from Schrecken by which the hamlet of Am Lech (see Rte. A) is reached in little more than 2 hrs. The traveller bound for Landeck may choose between the way by Krumbach, Steg, and the Kaiserjoch, or that by Am Lech and the Zürserthal described in Rte. A. The first is rather the shorter, the latter the easier route.

The shortest way from Schrecken to Oberstdorf is by Krumbach (1 hr.), and thence over the low pass called *Haldenwanger Eck* (6,070'), between the Biberkopf and the Widderstein, leading to the head of the Stillach glen. (Rte. A.)

Route C.

BREGENZ TO SONTHOFEN, BY THE BREGENZER WALD.

Those who love to follow unbeaten tracks, and to visit districts whose population has preserved its primitive usages unbroken amidst the changes that have altered so much else in Europe, can scarcely do better than devote a few days to rambling on foot through the district lying between the Bregenzer Ach and the Iller. With a good map and a knowledge of German, they will have no difficulty in finding their way from one place to another, but will often discover that the actual distance is very much greater than they would have inferred from the map. The traveller whose object is to reach the valley of the Iller, will have a choice amongst many different routes, most of which are practicable only for the pedestrian. He who prefers to travel in a wheeled vehicle may follow the road mentioned in the last rte., which leads from Alberschwendy to Lingenau. A good road leads thence to *Hüttisau* (2,962′) This stands on the flattened ridge separating the stream of the *Sübersbach*, or Süberach, from the *Bolgenach*. The district abounds in mineral springs, most of which contain alkaline salts associated with sulphur.

The carriage-road is carried NNW. from Hüttisau to Krumbach (not to be confounded with the village so named in the upper Lech Thal), whence it leads to the Oberstaufen station on the road from Augsburg to Lindau. Another road crosses the Bolgenach near its junction with the Weissach, and is carried NE. by Riefensberg to *Staufen*, a small town (good inn in the market-place), near the high road from Bregenz to Immenstadt. The pedestrian who has followed the above route as far as the valley of the *Weissach* may follow that stream to its source in a glen lying on the N. side of the ridge of the *Rindalpenhorn* (6,442′), and then choose between a track leading direct to Immenstadt, or another bearing to the rt. which will carry him to Seyfriedsberg (Rte. A), less than 1 hr. from Sonthofen. From Hüttisau the pedestrian may find a more direct way to Sonthofen by following one or other of the two branches of the Bolgenach which unite a short distance above the former village. The northern branch, flowing through the *Lechnerthal*, keeps a more uniform course; the more interesting route is by the S. branch, passing the village of *Balderschwang*. The upper portion of both valleys belongs to Bavaria. Instead of following the Bolgenach, the pedestrian bound for Algau may follow the glen of the Sübersbach to the highest village called *Sibratsgfäll*. The mountain ridge dividing the head of this glen from the valley of the Iller is accessible in many directions, but the most frequented track will lead him by the Starzlach torrent to Oberstdorf (Rte. A.)

In every part of the Bregenzer Wald the traveller is struck by the appearance of comfort and independence. Large, roomy, well-constructed houses, massive old-fashioned furniture, universal cleanliness, and almost universal friendliness, leave a distinct and pleasant recollection in the mind of the visitor.

Route D.

SONTHOFEN TO REUTTE IN THE LECH THAL.

	Stunden	Eng. miles
Hindelang	2	6
Schattwald	3	9
Höfen	2	6
Weissenbach	3½	10½
Reutte	2½	7½
	13	39

The passes described below connecting the Algau with the Lech Thal are traversed by a good road, which is, however, not supplied with post-horses.

'In 1861 a party paid 20 Bav. florins for a carriage and two horses by this route from Immenstadt to Reutte, besides 2½ fl. from Hindelang to Schattwald for extra horses, and 1 fl. for *Trinkgeld* to the driver.' [G.C.C.]

Distance from Immenstadt to Reutte about 15 stunden; time about 8 hrs. Taken in connection with the rly. from Lindau to Immenstadt, and the road from Reutte to Innsbruck described in § 42, this offers a very direct and agreeable route for travellers wishing to enter Tyrol from the Lake of Constance by a carriage-road.

At Sonthofen (Rte. A) the road to Reutte leaves the main valley of the Iller, and ascends a little S. of E. along the valley of the Ostrach to *Hindelang* (2,693′), a thriving village with a good inn (Adler), perhaps the best in this district. The upper valley of the *Ostrach*, which may be conveniently visited from hence, is further noticed below, as it offers to the mountaineer the most interesting route to the valley of the Lech. Another excursion to be recommended is that to the wild glen of the *Besondere Ach*, a torrent that joins the Ostrach about 3 m. from Hindelang. The head of the glen is enclosed by bold limestone peaks exceeding 7,000 ft. in height.

The high road soon after leaving Hindelang quits the Ostrach, and begins to ascend by zig-zags a rather steep hill, passing on the way the village of Obersdorf, not to be confused with Oberstdorf on the Iller (Rte. A). The ascent terminates at the summit of a ridge that divides the Ostrach valley from the head of the Wertach, a stream running N. to the plain of Bavaria. The pass, called *Vordere Joch* (3,765'), commands a fine view of the valley of the Ostrach and the glen of the Besondere Ach. A slight descent, and then another ascent, carries the road from the first to a second pass, called *Hintere Joch*, which marks the frontier between Bavaria and Tyrol. Looking eastward this overlooks a picturesque mountain basin, bright with green pastures and alpine flowers, backed by the bold limestone peaks of the *Schaffschroffen* (7,323'). There is a small establishment of mineral baths at *Schattwald* (Inn: Traube, good), a village standing in the upper part of the Vilsthal, whose torrent joins the Lech between Füssen and Reutte after a very circuitous course, so that the road ascends the valley instead of following the course of the stream. The next village is *Höfen* (3,657'), erroneously called *Tannheim* on Mayr's map, as that name belongs to the whole district from Schattwald to the sources of the Vils, of which this is the principal place. It has a tolerable inn.

Ascending southward from Höfen along a torrent, the traveller may visit the *Vilsalpensee*, a rather large mountain lake, whence a path leads to the W. over a pass to the Hintersteinerthal (mentioned below). Another mountain lake, called Fraualpsee, lies at a higher level SE. of the first. Both abound in trout. Near Höfen the road passes a cleft in the mountain, called Bognerberg, from whence at certain times strange sounds resembling moaning or howling are heard to issue, along with blasts of air, said to be violent enough to overturn passing vehicles. These phenomena are observed to precede the approach of bad weather, and are doubtless connected with barometric changes.

E. of Höfen the valley extends for some miles nearly at a dead level. To the N. is an opening through which a stream descends to join the Vils near Steinach, while another opening at the E. end of the plateau affords a passage for the road leading to Weissenbach. It would appear that the greater part of this space was once occupied by a lake, now partly filled up, and that it may have supplied streams that flowed alternately, or even simultaneously, through both the above openings, as well as to the main channel of the Vils. Part of the basin of this former lake is still occupied by the *Haldensee*, picturesquely placed at S. base of the Schaffschroffen, and well stocked with fish, like most of the lakes of this district.

The defile through which the Weissenbach torrent descends to join the Lech is called *Pass Gacht*. The scenery is very striking, especially at the junction of the Birkbach, where both torrents have cut deep clefts in the triassic rocks. The descent is rather steep, requiring caution in the driver, and the valley of the Lech is reached at the village of *Weissenbach*, with a fair country inn, 2½ stunden from

Reutte (Inns: Post, good and reasonable; Krone), further noticed in § 42.

[The pedestrian may diverge from the route above described, and enjoy some fine Alpine scenery by turning southward at Hindelang and following the glen of the Ostrach. In about 1½ hr. he reaches *Hinterstein*, the highest village in the glen, which, however, extends to the S. for many miles, bearing the name *Hintersteinerthal*. Above the village the torrent traverses a very deep cleft, called Eisenbrecherklamm, through which the path mounts to the upper level of the glen extending to the base of the Hoch Vogel, one of the highest peaks of this district. The traveller may rejoin the carriage-road at Höfen by a pass which will lead him by the Vilsalpensee (see above), or by following the torrent to the foot of the Hoch Vogel he may reach Forchach on the Lech (Rte. E) by a track which descends due E. along the Schwarzwasser torrent. On the west side of the glen is a pass leading nearly due W. to Oberstdorf, and another nearer to the Hoch Vogel, by which that village is reached through the Oythal. Both these passes are mentioned in Rte. A. The *Hoch Vogel* (8,501′) is said to be inaccessible from the Tyrolese side, but the summit is reached without much difficulty from the head of the Hintersteinerthal. Crampons (*Steigeisen*) are said to be requisite, as the way lies for ½ hr. over a steepish slope of névé.]

Route E.

FELDKIRCH TO REUTTE, BY THE WALSERTHAL AND LECH THAL.

	Stunden	Eng. miles
Satteins . . .	2½	7½
Thüringen . .	2½	7½
Blons . . .	1¾	5¼
Buchboden . .	3	9
Krumbach . .	5	12½
Lechleiten . .	1½	4½
Holzgau . . .	3	9
Elbigen Alp . .	2½	7½
Stanzach . . .	4	12
Weissenbach . .	2½	7½
Reutte . . .	2½	7½
	30¾	89¾

To the traveller looking out for an unfrequented route from Switzerland to Bavaria that here briefly described may be acceptable, although it presents no scenes of unusual interest. About two-thirds of the way are practicable in a wheeled vehicle, but on the whole it seems better suited for the pedestrian, who, if pressed for time, may accomplish the distance in three days, sleeping at Buchboden and at Elbigen Alp. It will be obvious that either the E. or W. portions of this rte. may be easily combined with a visit to the Algau or the Bregenzer Wald (see Rtes. A and B).

A short distance from *Feldkirch* (Inns: Post; Engel; both good), the traveller bound for the Walserthal quits the high road of the Arlberg (described § 34, Rte. A), and crosses to the rt. bank of the Ill, where the first considerable village is *Satteins*. This may be approached from Rankweil in the valley of the Rhine by a country road passing Göfis in much less time than by Feidkirch. From Satteins the road runs along the base of the mountains to *Bludesch*. The picturesque scenery of the Wallgau, or lower valley of the Ill, is enhanced by the remains of many old castles, and from time to time by fine views of the summits of the Rhætikon Alps to the S. The very ancient church of St. Nicholas, E. of Bludesch, deserves notice.

Thüringen (2,066′) stands at the opening of the Walserthal. A large cotton-mill worked by powerful water-wheels was established here by two Englishmen several years ago. The track leading through the *Walserthal* mounts gently along the rt. bank of the *Lutzbach* to *St. Gerold* (2,520′). Here is the site of an ancient monastery founded in the 10th century on the site of a hermitage, whither St. Gerold, a member of the ducal house of Saxony, retired from the world. In the church are shown the tombs of the saint and his two sons. N. of the village rises the *Hoch Görrach* (6,415′), and on the opposite side of the valley the *Frassen* (6,465′).

The next village to St. Gerold is *Blons*, opposite to which on the l. bank is *Raggal*, at the opening of the *Maruelthal*, a glen running deep into the mountain range that divides the Walserthal from the Klosterthal and terminating at the foot of the Rothewand.

The hamlets and solitary houses scattered through the upper end of the Walserthal all belong to the commune of Suntag. The chief village, rather more than 1 hr. from Blons, is 2,850 ft. above the sea, and beyond it the path ascends rapidly to *Buchboden*, where there is a poor inn. The mineral spring of Fontanella lies on the N. slope of the valley, and there is a track passing that way to the Argenthal (Rte. B) in the Bregenzer Wald. Nearer to the track are the baths of *Rothenbrunn* (4,232'), where a mountaineer might probably find night quarters. The most direct and easiest route from the Walserthal to the upper valley of the Lech is to cross the comparatively low pass at the head of the former valley leading to Schrecken at the head of the Bregenzer Ach (Rte. B). It is a walk ot but 1 hr. thence to *Krumbach* (5,481'), passing the Körbersee, which lies on the summit level dividing the Bregenzer Ach from the stream that descends by Warth, 1 hr. below Krumbach, to join the Lech below Lechleiten.

[The mountaineer may take a course which is probably more interesting than that usually followed by the natives of the Walserthal. Opposite Buchboden a glen opens due S., and leads to the N. base of the *Rothewand* (8,847'), on which peak the botanist may find *Campanula cenisia*. Then turning eastward he may traverse a pass lying between the Rothewand and the Hirschenberg, and thus descend into the head of the Lech valley, and follow the course of that torrent to Lech (Rte. A)]

Besides Warth and *Lechleiten* several hamlets are passed before reaching the fair inn at *Steg* (3.725'). Thence a road practicable for light carriages extends down the valley till it joins the high road at Weissenbach. The first place of any importance is *Holzgau*, formerly the seat of a considerable trade in timber, now much reduced by the felling of the forests. There are several separate groups of houses, each having a country inn, of which the best is that at *Höhenbach* (3,695')—pronounced Hechenbach. Here diverges a track noticed in Rte. A, which leads, in 7 hrs., to Oberstdorf in Algau over the *Mädelejoch*.

Below Holzgau the road crosses to the S. bank of the Lech at a point where the mountains approach the stream on either hand. This marks the boundary between the portion of the valley locally called Oberlechthal, extending hence to the source of the river, and that called Unterlechthal, which terminates at Stanzach. The lower part of the valley, formerly belonging to Bavaria, is not included by the Tyrolese in the second division; but in this work the designation Lech Thal, or valley of the Lech, includes the entire course of the river as far as Füssen, where it enters the Bavarian plain. The passes leading from the Lech Thal to the valley of the Inn are noticed in § 42.

After passing *Stockach* the road returns to the l. bank of the Lech at *Lend*, lying opposite to the opening of the Lendthal. Here the traveller enters the most thriving portion of the valley, passing many large and comfortable houses belonging to persons who have carried back to their Alpine home the savings accumulated by trade in various parts of Europe and America. The centre of this district is

Elbigen Alp (3,515'), with a very good country inn, a handsome church, and au ancient chapel of St. Martin, built when the adjoining slopes were merely the summer resort of herdsmen from the lower valley. About 1 hr. below Elbigen Alp is *Unterhofen*, where there also is an inn supplying excellent beer. Here the road crosses to the rt. bank close to the opening of the Gramaiselthal. About 1½ hr. farther another glen is passed on the rt. hand, called Pfafflarthal. Both are noticed in § 42. After passing *Elmen*, the considerable torrent of the

Hornbach is seen to descend into the main valley from the W. Through the very picturesque *Hornthal* a path leads over a steep and rough pass to Oberstdorf in Algau, through the Oythal (Rte. A).

The portion of the Lech Thal extending hence to Weissenbach has been made unsightly by the devastations of the torrent, that has spread gravel and slime over the floor of the valley. By reducing the extend of land available for tillage, the same cause has driven a portion of the population to wander abroad in search of a livelihood.

Leaving the village of *Hornbach* on the opposite bank, the road keeps by the rt. side of the Lech through *Stanzach* (3,007′) and Forchach, returning to the l. bank at *Weissenbach*, where the road from Immenstadt descends through the Pass Gacht into the Lech valley. 2½ hrs. farther is Reutte (§ 42, Rte. B.). A post carriage plies three times a week between that place and Steg.

SECTION 42.

ZUGSPITZ DISTRICT.

THE central portion of the Suabian Alps described in this section is clearly defined by the rivers Lech and Isar to the W. and E., and by the valley of the Inn and the Stanzerthal to the S. It has been mentioned in the last section that the head waters of the Lech flow from a valley which approaches very near to the Arlberg Pass, and the track leading from Stuben to Am Lech was fixed upon as the boundary between this and the preceding district. In the same way the Upper Valley of the Isar extends close to the crest of the mountains overlooking the valley of the Inn, and within 2 or 3 m. of the town of Innsbruck the herdsman looks down on the one side upon the capital of Tyrol, and on the other upon one of the sources of the Isar. The natural break, however, in the range enclosing the valley of the Inn is marked by the Pass of Seefeld, where the road from Mittenwald to Zirl affords the easiest and most frequented route from western Bavaria to the Tyrol.

The mountains included within the limits above described are formed exclusively of jurassic and triassic rocks, unless it should hereafter appear that a portion of the latter ought to be referred to the Verrucano, while another may possibly belong to the Infra-Lias of the Lombard geologists. The most considerable mass is that dividing the basin of the Inn from that of the Lech, and extending with no considerable depression from the Arlberg Pass to Lermoos. At least two summits of this range— the Muttekopf (9,077′) and the Stanzerkopf (9,041′)—exceed 9,000 ft. in height. The highest point, however, of the present district is the Zugspitz (9,716′), which crowns the comparatively small and isolated group of the Wetterstein Gebirge, SW. of Partenkirch, forming the boundary of Tyrol and Bavaria.

In spite of the irregularity which usually characterises the orography of limestone districts, the recurrence of ridges and valley running E. and W. in a direction transverse to that of the general drainage of the district will not escape the notice of the physical geologist. The Geisthal, the middle portion of the Loisachthal, the Graswangthal, and the ridges by which these are enclosed, not to mention many minor glens, may serve to illustrate this observation.

To the sportsman as well as the naturalist this district offers many attractions. Chamois as well as trout abound; but in many places, especially in Bavaria, they are preserved, and the stranger must obtain permission which it is not difficult to procure.

Good accommodation is found at many points in this district, especially at Partenkirch and Walchensee, and the mountaineer need seldom be at a loss for tolerable quarters in the inns which are found in every village.

The larger part of this district belongs to Tyrol, but the northern portion, lying in Bavaria, includes a great deal of charming scenery, and the ascent of the Zugspitz is an expedition which rivals in interest, as well as in difficulty, that of many higher and more renowned peaks.

Although the great valley of the Inn, marking the natural boundary between northern and central Tyrol, as well as the geological limit between the crystalline and sedimentary rocks, merely forms the southern limit of the district here described, it seems convenient to insert in this place the description of the tract lying between Landeck and Innsbruck. The upper course of the same river, lying chiefly in Switzerland, has been described in § 36 in connection with the central range of the Rhætian Alps.

The distances charged in posting, and here adopted for the main roads, appear to the writer and to other travellers to be exaggerated.

Route A.

LANDECK TO INNSBRUCK, BY THE INN-THAL.

	Austrian miles	Eng. miles
Imst	3	14
Silz	2½	11¾
Telfs . . .	1¾	8¼
Zirl	2¼	10½
Innsbruck . .	1¾	8¼
	11¼	52¾

The traveller who has entered the Tyrol by the Arlberg road, described in § 34, Rte. A, or reached Landeck on the Inn by the Routes A or B, described in the last section, has before him one of the main valleys which give its characteristic features to the orography of Tyrol. Having maintained a very direct course from its source near the Maloya Pass to Prutz (§ 36, Rte. A.), the Inn is there turned aside towards the NW., and after running for some miles through a tortuous channel, enters at Landeck the depression which marks the division between the crystalline rocks of the central chain and the sedimentary rocks of the Suabian Alps. The valley of the Inn is traversed by a high road connecting Innsbruck with the E. of Switzerland by the Arlberg Pass, and with the valley of the Adige by the defile of Finstermünz.

Besides the diligence (Eilwagen) carrying passengers in 25 hrs. from Bregenz to Innsbruck, a country omnibus (Stellwagen) starts daily from Landeck for Innsbruck. The fare is, or was, only 2 flor.

The post-road to Innsbruck follows the rt. bank of the Inn through a rich level tract abounding in fruit-trees till the river is crossed by a wooden bridge at *Zams* (2,725′), a thriving village with a cotton-mill, schools, and a convent of sisters of charity. About ½ m. to the l. is the hamlet of Letz on the *Letzerbach*, a torrent descending from the Matrioljoch (Rte. C). There is a curious waterfall here, approached only by permission of a miller, whose water-wheel is turned by the torrent, and who receives a small fee from visitors. The water descends with extreme rapidity, but without breaking into foam, through a steep cleft into a deep hollow caldron, where it boils with extreme fury. Parallel to the Letzerbach there formerly existed a strong wall with high towers, said to have been intended to resist the incursions of the Swiss into the valley of the Inn. The lower tower has fallen into the river, and another has been converted into a dwelling-house. The mountains in this part of the main valley descend in very bold and varied forms to the banks of the Inn. On the rt. bank the castle of *Kronburg* crowns a sharp pyramidal summit, and in the background rises the bold peak of the Tschürgant beyond Imst. As the road passes under the rock of Kronburg the valley is narrowed to a mere defile, but it opens somewhat at the hamlet of *Starkenbach*, where a track mounts through a glen of the same name leading to the Lech Thal (Rte. C).

[The pedestrian not overpressed for time may reach this point in the valley by a more interesting way than the high road if he will mount from Zams by the hamlets of Revenal and Christ to the pilgrimage church of Kronburg, standing with a few houses and a country inn on the saddle which connects the bold rock above mentioned with the mass of the Venetberg. The castle may be reached in ¼ hr. from the church. From the same point the traveller may descend in ½ hr. and rejoin the post-road at Starkenbach.]

Beyond *Mils* the road ascends along a steep wall of arenaceous (triassic?) limestone, which at some points has been carved into singular forms by the action of the elements. This was one among many points in the valley where the French troops suffered severely in 1809. The road turns somewhat away from the river at the opening of the Gurglthal, where stands the prosperous village of

Imst (Inn: Post, good, the best in the Oberinnthal), 2,696 ft. above the sea. The position of the village, on rising ground overlooking the junction of the Gurgl torrent with the Inn, is very picturesque. To the NW. rises the *Muttekopf* (9,077'), probably the highest summit of the range on the S. side of the Lech Thal, easy of access, and from its central position commanding an admirable panoramic view. In the background of the Gurglthal are seen the pale spectral form of the Sonnenspitze, one of the highest peaks of the limestone range of the Mundistock. Immediately E. of the village is the bold summit of the *Tschürgant* (7,545'), terminating the ridge which divides the Gurglthal from the main valley. The ascent from this side is said to be steep, but from its position the mountain ought to command a fine view.

The most conspicuous summit on the opposite side of the Inn is the *Wildgratkogel* (9,751'), dividing the lower part of the Pitzthal (§ 48, Rte. F.) from the Œtzthal. A short walk to the Gunglgrün may be recommended to those who have a couple of hours to spare at Imst. The hamlet commands a charming view, and the little inn is resorted to for breakfast on account of its excellent milk and cream.

Imst suffered cruelly in 1822 from a fire which destroyed the church and 216 houses. The trade in canary-birds, once very considerable, and extending over a great part of Europe, has all but completely disappeared, but new branches of industry have taken its place.

By a very slight detour the traveller bound for Innsbruck may follow the road to Nassereit by the Gurglthal, and rejoin the Innthal at Telfs; but the diligence follows the main road along the Inn valley, which is equally interesting. See Routes B. and F. About 1 m. from the village a little chapel marks the spot where that excellent mountaineer and naturalist, the late King of Saxony, was killed by being thrown from his carriage when on his way to explore the Pitzthal.

After passing Karres the high road bends to the l., and winds round the base of the Tschürgant, when a vast tract of debris and gravel spread over the floor of the valley announces the junction of the important torrent from the Œtzthal with the Inn. It is evident that at some former period, or possibly at recurring intervals, the masses borne down from the central range either by water or ice have sufficed to throw a dam across the main valley and hold back the stream of the Inn until this has been able to clear away the barrier. From one part of the road the traveller gains a fine vista along the valley of the Inn, and the contrast between the limestone peaks of the Hohe Mundi and the Solstein to the N. and the crystalline rocks forming the mountain range on the S. of the valley must strike even the least observant eye.

The Œtzthal, the most important of the tributary glens of the Inn valley, is described in § 48, Rte. B. The rough char-road leading to Œtz and Umhausen crosses the Inn some way above its junction with the torrent from the Oetzthal, and follows a track that passes the village of Roppen. In approaching the Oetzthal from Innsbruck, the traveller turns aside to the l. 3 m. below the junction, at *Haimingen*. Here the post-road from Landeck crosses to the l. bank of the Inn, and then passes beneath a rocky eminence planted with fine lime trees, on which stands the venerable Castle of Petersberg, better known in history as the Welfenburg. Its story is that of the princely and noble houses that have in turn ruled this portion of the valley. It was in turn the birthplace and the prison of Margaret Maultasch, Countess of Tyrol, the wife of John, King of Bohemia; and after passing through many changes of ownership has descended to Count Wolkenstein, whose possessions are found in many other parts of the Tyrolese Alps.

Silz (Inns: Post, good; and several others) has an air of prosperity and comfort that is common to most of the villages of the valley. A little farther on is Mötz, where the Inn becomes navigable for barges and timber-rafts. Keeping to the rt. bank the road soon reaches *Stambs*, where a remarkable monument of early ecclesiastical architecture well deserves a passing visit. The present monastery and church occupy the site of a convent founded in 1271 by Elizabeth, the mother of Conradin, the last of the house of Hohenstaufen. She did not long survive the murder of her son, and was buried here in 1273. Enriched by subsequent rulers of the Tyrol the building was enlarged, and became a stately Cistercian monastery, where Maximilian I. often resorted. Great hunting parties issued from the oak woods that surrounded the monastery, and on the open lawn before the building he received the ambassadors from Sultan Bajazet, when they came to demand the hand of his sister Kunigund as Christian wife of the would-be Christian successor of the Caliphs. The present monastery and church are comparatively modern, the former building having been destroyed by fire in 1593; but some curious ancient paintings were preserved, especially an altar-piece executed in the 14th century by the then abbot. The chief object of interest is the crypt, wherein are preserved the monuments of twenty-nine royal and princely personages, whose remains are here preserved. These are divided into three compartments, corresponding to as many remarkable periods of Tyrolese history. In the first compartment are the monuments of the last male members of the line of the Counts of Tyrol and Goritz, whose bodies were removed hither from the Castle of Tyrol, the cradle of their race. Along with them lies the foundress, Elizabeth, wife of the Emperor Conrad IV. In the second compartment are preserved the remains of Frederick IV., surnamed of the Empty Purse, near to the Alpine retreat in the Œtzthal where he lay long concealed from his enemies—not far from Landeck where, disguised as a minstrel, he stirred the brave hearts of the Tyrolese by the story of the wrongs and misfortunes of himself, their rightful prince—in the midst of the Inn valley where he achieved the final triumph over his powerful foes. Beside him are the tombs of his two wives and four children. In the third compartment rest the bodies of Sigismund and his wife

Eleanor of Scotland, and of seven princes of his house.

The next village to Stambs is *Rietz*, standing beside a destructive torrent that descends from the *Hocheder Spitz* (9,157'), a peak commanding a fine panoramic view. Soon after the road returns to the l. bank of the Inn, and reaches *Telfs*, a large village (2,026') with a good country inn. As in most places in Tyrol, the church deserves a visit. It contains frescoes of more than average merit, by Zoller, a native of this village, which gave birth also to the well-known painter Joseph Schöpf, and to some sculptors of local repute. The position of Telfs at the S. base of the *Hohe Mundi* is very picturesque. The best view of the neighbourhood is gained from the Calvarienberg. At Telfs the road from Kempten by Füssen and Nassereit (Rte. B) enters the valley of the Inn, and travellers from Imst who have made the slight detour by the latter village here rejoin the main road.

The post-road from Telfs to Innsbruck keeps all the way by the l. bank of the Inn, but there is a tolerable country road by the rt. bank, called the Salzstrasse. The shortest way for the pedestrian coming from Imst is to follow the latter as far as Zirl, turning off from the main road at the bridge above Telfs, and avoiding the latter village.

Following the high road, we leave on the l., near Telfs, the hamlet of Brand, whence a track is carried over a low pass, called Böden, in the range of the Mundistock, and leads in 1½ hr. to the Leutasch Glen (Rte. F). The road after traversing an open space in the main valley is obliged to cling to the rocks at a point where the river flows at the foot of the mountain. Then follows the hamlet of Platten, whence a rough cart-track passes by Mösern to Seefeld on the high road from Zirl to Mittenwald.

Zirl (Inns: Post; Löwe) is a village 2,039 ft. above the sea, ever since Roman times a centre of considerable traffic. The deep gap between the range of the Mundistock and the Solstein (forming the natural division between the two mountain groups described in this and the following sections) opens the portal through which the highway to Bavaria, described in Rte. D, here leaves the valley of the Inn.

Near to Zirl are two ancient castles, both hunting-seats of that inveterate sportsman the Emperor Maximilian. Northward near the Seefeld road is the Fragenstein, a picturesque object as seen from the village. Eastward is the Martinsbühel, on an eminence overlooking the river, and commanding a noble view through the main valley. It is said to occupy the site of a Roman fort. Roman milestones marked the distances on the road to Seefeld; one of them is preserved in the Museum at Innsbruck. In Tyrol the name *Ober Innthal* is applied to the portion of the valley of the Inn lying between Landeck and the foot of the Martinswand below Zirl, while the tract extending thence to Kufstein is called *Unter Innthal*.

The ascent of the *Solstein* (8,649') is an excursion much recommended, the mountain being more easy of access from this side than from Innsbruck. The shortest and least laborious way is by the Galtalp, but a more interesting route is said to be by a path turning off from the high road to Innsbruck at a wayside inn, called Kranebitten. After climbing, partly by ladders, through a singular cleft called Schwefelloch, the traveller reaches the pastures of the Zirler Alp, where he may pass the night. The view from the top is remarkable. On every side a girdle of peaks, with the contrast between the rugged limestone rocks of the Suabian Alps, and the snow-covered pyramidal summits of the Œtzthal, Stubay, and Zillerthal groups. Almost at the traveller's feet he peers down into the streets of the city of Innsbruck, and overlooks the busy valley of the Inn, the flow of the river, and the traffic on its roads, for a distance of fully fifty miles.

Nearly opposite to Zirl is the opening of the Selrainerthal, described in § 49, Rte. B.

The object which chiefly arrests the

notice of the stranger going from Zirl to Innsbruck is the *Martinswand*. This name is given to the almost vertical precipice at the base of the Solstein, which supports the pastures of the Zirler Alp. The narrow space through which the river passes at the foot of the rocks, while the road is carried along their lower ledges, has naturally been selected as a stronghold by the natives of the valley in their heroic struggles against French or Bavarian invaders; but the spot is best remembered as the scene of a well-known adventure in the life of the Emperor Maximilian. Losing his footing somewhere near the summit of the rock, which rises 1,832 ft. above the Inn, the rash sportsman was sliding or rolling down the fearful precipice, when some projecting bush or rock arrested his fall, and enabled him to hold on with the energy of desperation just above the point where the declivity becomes a mere wall, inaccessible even to the foot of the chamois. There he hung, Holy Roman Emperor as he was, and though a crowd of faithful subjects gathered at the foot of the precipice, no help was to be had. His strength was fast ebbing away, prayers were put up as for a man upon his death-bed, when a shout was heard from above; a human figure was seen to descend, to pass along ledges so narrow as to be scarcely perceptible from below, till it approached the almost despairing monarch. An angel! an angel! was the cry among the crowd, and even now the belief lingers among the people of the valley. It was a hunter named Zips, whose skill and courage as a cragsman saved the Emperor in his utmost peril. A steady head and strong arm will not only carry a man over many a spot that seems inaccessible, but enable him to give essential aid to a companion not absolutely rendered helpless by weakness or nervousness. The stout heart of Maximilian was not likely to give way at such a time, and it is certain that he escaped unhurt. An uncertain tradition affirms that the bold hunter was rewarded by a patent of nobility.

with the title of Hollauer von Hohenfelsen, in memory of the loud halloo by which he announced his approach when he had perceived the Emperor's perilous case. There is more direct evidence for the fact that the latter paid to his preserver an annual pension of 16 florins.

On the way to Innsbruck the geologist will not fail to remark the vast dimensions of the beds of gravel and rolled blocks which must have extended across the valley of the Inn at a period geologically very recent. It is an agreeable drive of about 8 m. from Zirl to

INNSBRUCK (Inn : Œsterreichischer Hof, Goldene Sonne, both tolerably good, the former ranking highest; H. de l'Europe, new, close to the railway station; Goldener Adler, old-fashioned; Hirsch; Löwe; Goldene Stern on the l. bank of the Inn, good and reasonable). The capital of Tyrol can boast of a site such as few cities in Europe can rival. It does not, indeed, like Turin, command a horizon girdled by the range of the snowy Alps, nor are its walls, like those of Geneva or Lucerne, washed by the azure waters of a lake, in which the distant background is mirrored; its position may rather be compared with that of Coire, or Villach, or Trent, or Aosta, or Susa— all, like it, lying in a broad Alpine valley, wherein the rich vegetation of the lower zone is brought into close contrast with the sternness of the impending mountain summits. In the height and boldness of the surrounding peaks Innsbruck surpasses all save the two last-named rivals; but here the prevailing hues are different, and a brighter verdure replaces the rich brown and delicate grey tints that predominate in the Italian valleys. Although essentially a German city, it has borrowed something in the style of its construction from Italy, with which it has from the earliest times held constant intercourse by the adjoining pass of the Brenner, the lowest and easiest of access of all the passes of the Alps. The ancient bridge, 1,882 ft. above the sea, whence the city takes its name, and the modern suspension bridge leading to Mühlau, command excellent views of the surrounding mountains, amongst which the Solstein is pre-eminent.

Innsbruck has most of the appurtenances of a provincial capital—a palace, with adjoining public gardens, a university, a new theatre, an academy of music, a casino, with a reading-room liberally opened to strangers, and a museum, called the Ferdinandeum. The latter institution, founded by Count Chotek, well deserves a visit. It includes, along with a few good pictures by ancient masters, some creditable specimens of native art by little-known Tyrolese painters. Some antiquities, relics of Hofer, and other objects of local interest are also to be seen, as well as specimens of native manufacture and handicraft. There is also a good collection of Tyrolese minerals and fossils, and a pretty complete herbarium of the local flora. In connection with the museum, a periodical publication (called Ferdinandeum) has contained much interesting information, chiefly connected with natural science.

Among the numerous churches the first in rank—the Pfarrkirche—contains little of interest except a small picture by Lucas Cranach, which is regarded with peculiar veneration. It is inserted, as in a frame, in a larger modern work by Schöpf.

The great object of artistic and antiquarian attraction is the tomb of the Emperor Maximilian, contained in the Franciscan church of the Holy Cross. This, along with the other works mentioned below, amply deserves the careful examination of every stranger visiting the city. The architecture of the church presents a feeble combination of gothic and renaissance styles, characteristic of the middle of the 16th century, when it was built. Attention is at once concentrated upon the massive marble sarcophagus standing in the centre of the church, and supporting the kneeling figure of Maximilian in bronze, with the face turned towards the high altar. The sides of the sarcophagus are adorned with twenty-four elaborate bas-reliefs in white marble.

All but four of these were the work of Alexander Collin, of Mechlin, who here proved his claim to rank in the highest class of mediæval sculptors. He is said to have received for each of them the moderate sum of 240 florins. As the visitor must pay a small fee to the guardian who removes the screens with which they are habitually covered, it is unnecessary to enumerate the subjects represented in these remarkable works, which combine extreme, almost excessive, accuracy and delicacy of detail with a fine sense of pictorial effect. It may strike the visitor as strange, that in the series of marble pictures designed to illustrate the life of Maximilian, his romantic adventure on the Martinswand, almost within sight of the church, should have been omitted. Of less artistic merit, but imposing from their colossal size and the poetical idea which they embody, are the bronze figures ranged around the church as guardians of the tomb of the deceased Emperor. In accordance with the ideas of the time, there stand here, along with the relatives and immediate ancestors of the Imperial dead, several of the semi-legendary heroes whose names were held in reverence in the popular faith of the middle ages. Beginning at the left hand, and going round the nave, the statues hold the following order:—
1. Joanna of Spain, mother of Charles V. 2. Ferdinand of Aragon, father of the last. 3. Kunigund, sister of Maximilian. 4. Eleanor of Portugal, his mother. 5. Mary of Burgundy, his first wife, daughter of Charles the Bold. 6. Elizabeth, wife of Albert II. 7. Godfrey of Bouillon. 8. Albert I., Duke of Austria. 9. Frederick (surnamed 'With the Empty Purse'), Count of Tyrol. 10. Leopold III., Duke of Austria. 11. Rudolph, Count of Habsburg, grandfather of Rudolph I., the first Emperor of that line. 12. St. Leopold. 13. Frederick III., father of Maximilian. 14. Albert II. 15. Philip the Good,

of Burgundy. 16. Charles the Bold, son of the last. 17. Cymburgis, wife of Ernest, the Iron-hearted. 18. Margaret, daughter of Maximilian. 19. Bianca Maria Sforza, his second wife. 20. Sigismund, Count of Tyrol. 21. King Arthur of England. 22. Theobert, Duke of Burgundy. 23. Ernest the Iron-hearted. 24. Theodoric, King of the Ostrogoths. 25. Albert the Wise, Duke of Austria. 26. Rudolph I. of Habsburg. 27. Philip I. of Spain, son of Maximilian. 28. Clovis, the first Christian King of the Franks.

Attached to the church is the *Silver Chapel*—so called from a silver image of the Madonna—designed by its founder, Ferdinand II., as a mausoleum for himself and his wife, the beautiful Philippina Welser. The reclining figures seen on the two monuments, and four bas-reliefs on that of Ferdinand, scarcely inferior to those on the tomb of Maximilian, have been attributed to the same artist, Alexander Collin. Those on the tomb of Philippina are by an inferior hand. In the same building with these memorials of the great is the modern monument to the peasant-hero of Tyrol, Andreas Hofer. His bones were removed hither from Mantua in 1823, and in 1834 a marble statue in the national costume was placed over the grave.

There are two or three unusually good booksellers' shops in Innsbruck. The usual hire paid for a carriage to Amras is 2 florins; to the Martinswand, 4½ fl.; to Schönberg on the Brenner road, 4½ fl.

Innsbruck is the centre towards which converge several of the most important lines of communication of SW. Germany. By the railway to Kufstein and Rosenheim, it is connected on the one hand with Salzburg, on the other with Munich, and that over the Brenner brings it within a few hours of Verona. The road above described leads by Landeck and the Arlberg to Switzerland, or by the Finstermünz and the Stelvio Pass to the Lake of Como, while the branch road from Telfs to Füssen (see next Rte.) is the way to Augsburg, and that from Zirl to Partenkirch (Rte. D) is the most direct way to Munich.

The two favourite mountain excursions from Innsbruck are the ascent of the Patscherkogl (§ 50, Rte. D) and that of the Solstein, noticed above. Those who cannot climb heights should not miss the view from the *Lanser Kopf* (3,100'), about 5 m. from the town. A carriage road leads close to the top.

The neighbourhood of Innsbruck and the valley of the Inn, both above and below the city, are hallowed by the recollection of the heroic struggle of the Tyrolese against the united forces of France and Bavaria in 1809. The limits of this work do not admit of an outline of the eventful struggles of that year, when the Tyrolese, three times victorious against overwhelming odds, were finally forced to submit, and the triumph of the first Napoleon was worthily completed by shooting in cold blood the peasant chief who had so often overcome his armies in the field.

For the favourite excursion to Amras see § 43, Rte. B.

Route B.

AUGSBURG, OR LINDAU, TO INNSBRUCK, BY FÜSSEN AND LERMOOS.

	Post miles	Eng. miles
Biesenhofen . .	8½	39½
Füssen . . .	4½	20¾
Reutte . . .	2	9¼
Lermoos . . .	3	14½
Nassereit . . .	2	9¼
Mieningen . .	2	9½
Telfs . . .	1½	7
Innsbruck . .	4	18¾
	27½	128½

Railway to Biesenhofen; post-road thence to Innsbruck; Bavarian miles as far as Füssen, Austrian miles thence to Innsbruck.

Travellers approaching this district from Lindau, on the Lake of Constance, leave the railway at Kempten, and may shorten the way by taking the direct road from Nesselwang to Reutte by Vils, avoiding the picturesque town of Füssen, and saving about one German mile. Reckoning by time, the shortest way from Augsburg to Innsbruck is to follow the railway by Munich, Rosenheim and Kufstein, but the present rte. is in all respects the more interesting. In approaching Innsbruck from the Lake of Constance, the traveller may choose between this way and that by Immenstadt and Reutte (§ 41, Rte. D), either being considerably shorter than that of the Arlberg. A diligence leaves the Biesenhofen station daily in the afternoon, and reaches Füssen in 4½ hrs. A more interesting road is that by

Kempten (Inns: Krone; Strauss; Hase), an ancient town (Campodunum of the Romans), on the Iller, which from hence to Ulm becomes navigable for barges. The Protestant lower town and the Catholic upper town are still divided by sectarian distinctions. Amidst a pleasing country, dotted with many scattered farm-houses, and enjoying at intervals fine views of the Alps, the road to Füssen reaches

Nesselwang (2,829'), a market-town at the N. foot of the *Edelsberg* (5,362'), an outlier of the limestone range between the Lech and the Iller, commanding a fine distant view. About 2 m. from Nesselwang the traveller will remark with surprise that the sluggish stream of the *Faule Ache*, instead of following the general direction of the drainage of this district, flows SE. from the plain towards the mountains.

[Those who are pressed for time may here follow a road that pursues a nearly straight course to ESE., passing the small Tyrolese town of *Vils* (2,551'), and entering the valley of the Lech between Füssen and Reutte. The neighbourhood of Vils is rich in rare plants. *Epipogium Gmelini* (found on the Salober) and *Crepis succisæfolia* may be specially noticed. The course of the Vils torrent may be studied with interest by those geologists who are disposed to attribute the formation of all Alpine valleys to excavation by the action of the elements, and who would exclude the agency of those mechanical forces that have produced the mountain ranges by which they are enclosed. Flowing from the Haldensee (§ 41, Rte. D), and swollen by the torrents from the adjoining mountains, the Vils proceeds at first about due W. to Schattwald; turning to N., and then to NE., it escapes through a mere cleft in the mountains, and descends to Pfronten on the road from Nesselwang to Vils. Here there is a broad opening to the N., and the natural course of a stream whether of water or ice would be towards the plain of Bavaria. Instead of taking that direction, the Vils, after meeting the Faule Ach (mentioned above), turns SE., and ultimately joins the Lech at a point not 5 m. distant from its source.]

The road from Nesselwang to Füssen keeps to the Bavarian side of the frontier, passes N. of the pretty lake of Weissensee, commanding fine Alpine views, in which the Schaffschroffen—here called Aggenstein—is conspicuous (§ 41, Rte. D), and then approaching the Lech, reaches

Füssen (Inns: Post; Sonne; Mohr), one of the most attractive little towns of S. Germany, 2,631 ft. above the sea, whose castle, towers, and spires, rising above the ancient walls, form from every side a charming picture. The castle deserves a visit for the sake of the building and for the view from the Storks Tower. The antiquarian may find some interest in the abbey and church of St. Magnus (here called Mang), which was founded in the 8th century, but rebuilt about 1701, so that little remains of the original structure.

No stranger visiting Füssen will fail to make a short excursion to *Hohen Schwangau* (2,832'). This ancient castle, perched above a little Alpine lake surrounded by high mountains, only 3 m. from Füssen, was restored to more than its ancient splendour by the late King Max, when Crown-Prince of Bavaria, who summoned to his aid the best artists of Germany in order to achieve an idealised reproduction of the mediæval feudal castle. The fresco paintings are of considerable merit, and the subjects for the most part happily chosen. To describe the castle in detail does not enter into the plan of this work. It is approached by a good carriage-road, or by a shorter footpath. The traveller bound for Reutte may obtain permission from the Bezirkvorsteher to pass by the Königsstrasse, or King's drive. The guides lead strangers by a circuit of about 2 hrs. to several picturesque sites, returning to the village of Hohen Schwangau, where there is an inn. A more arduous excursion is the ascent of the Säuling (6,673′), standing on the frontier of Tyrol and Bavaria, and commanding the best panoramic view of this beautiful district.

Between Reutte and Füssen the Lech escapes into the plain through a series of clefts cut through relatively high transverse ridges of limestone, instead of flowing through the broad portals by which it might have passed from Reutte to Nesselwang by Pfronten, or to Schwangau by the Alp See.

After crossing the bridge opposite Füssen, the road to Reutte turns SSW., towards the cleft through which issue the foaming waters of the Lech. This soon becomes so narrow that the road is forced to turn aside from the stream. Having reached the summit of a low col, the traveller should follow a path to the rt., which leads in a few minutes to the *Lech Fall*, or cataract of the Lech. The river is seen from above as it rushes amid huge blocks into the narrow chasm below. The picturesque effect is damaged by a mill-dam thrown across the stream just above the fall. A short distance farther is the Austrian frontier. After travelling WSW. for some distance, a very abrupt turn of the valley shows the Lech flowing from SE., and the road crosses it by a fine bridge. Here the road from Vils (mentioned above) joins that leading to Reutte. The prevalence of minor ridges and depressions running east and west, parallel to the strike of the strata, and transverse to the general direction of the drainage, which is more or less evident throughout this district, is especially obvious in this part of the valley of the Lech.

There are two roads, one following each bank of the stream, and the valley opens out into the level plain (bed of an ancient lake), where stands, 2,935 ft. above the sea, the pretty market-town of *Reutte* (Inns: Post, good and reasonable; Krone). On the opposite side of the Lech is the village of *Aschau* (also called Am Lech), a place of high antiquity, already in existence at the time of Pepin the Less, who granted it to St. Magnus, the apostle of this district. It long preserved its own peculiar laws and customs, some of which may not be yet altogether obsolete. For the passes over the mountains S. of the Lech, see next Rte.

The high road to Innsbruck leaves the valley of the Lech at Reutte, passing on the W. side of a conspicuous isolated mountain called Tauern—a name with which the reader will become familiar in E. Tyrol. After passing the Baths of Krekelmoos the road reaches the Klause, a defile overlooked by the picturesque ruins of the castle of *Ehrenberg*, whose importance as a frontier fortress guarding this entrance into Tyrol endured from the time of the Romans until the year 1800, when it was blown up by the French. After ascending (on foot) through the Klause it is worth while to turn aside and enjoy the view from the castle. The new road avoids the Klause, and ascends through a wooded hollow, until it emerges upon the upland plain near the first Tyrolese village of *Heiterwang* (3,319′), with a fair country inn (Hirsch).

[Although the way just described is not devoid of interest, there is a more attractive route by which the pedestrian may reach Heiterwang, passing by the N. and E. sides of the Tauern. After traversing the village of Breitenwang, with a very ancient church, a path mounts the hill called Rossrücken, commanding a fine view of the surrounding mountains, and then descends a little to enter the wooded glen called Achenthal —a name constantly recurring in the Tyrolese Alps—through which the torrent from the Plan See issues to join the Lech about 2½ m. below Reutte. Advancing through the forest, the traveller hears the deep roar of a waterfall, and before long reaches the four cascades collectively called *Stuibenfall* or Staübifall. The lowest of these is the finest, being over 90 ft. in height. A slight wooden bridge and a footpath enable visitors to gain the most favourable points for viewing the fall. The path continues by the l. bank of the torrent, before long joining a rough cart-track from Reutte, and in about 1 hr. from the lower fall the traveller reaches the shore of the *Plan See*, a rather large lake 3,244 ft. above the sea, lying in the midst of the mountains, which here have a singularly wild and deserted aspect. Although there is no trace of inhabitants, the cart-track is carried along the N. side of the lake and then mounts NE. through a short glen called *Ammerwald*, crosses a col, and descends into the Upper Ammergau in Bavaria (Rte. E). Another track running due E. leads to Garmisch and Partenkirch (Rte. F). The traveller intending to rejoin the road to Innsbruck will turn to the rt. on reaching the shore of the Plan See, and keep along a very rough path which follows the water's edge at the base of the Tauern. The western extremity of the lake, partly filled up with the debris poured in by a torrent, extends close to Heiterwang, where we rejoin the highroad.]

The peaked mountain S. of Heiterwang, called *Thaneller*, is said to command the finest view of this neighbourhood. The summit may be reached from Heiterwang or Büchlbach.

The road from Heiterwang to Lermoos lies throughout in the same line of valley, although the water runs at one end towards the Plan See and the Lech, and at the other to the Loisach and the Isar. This disconnection between the disposition of the valleys and the direction of the drainage is of common occurrence in the Eastern Alps.

At *Büchlbach* (3,505'), a path mounts westward through a lateral glen to *Bärwang* (4,465'), and then crosses the ridge separating this from the Rothlechthal (Rte. C). Before reaching Lähn the watershed between the Lech and Loisach is passed, and a fine view opens to E. and SE. of the lofty mountains that rise behind the green plain of Lermoos.

The *Wettersteingebirge*, culminating in the Zugspitz (9,716') are the higher and more massive, but the parallel range of the *Mieminger gebirge*—also called *Mundistock*—which attains 8,856 ft. in height, is broken into more varied and picturesque forms.

Lähn (3,693') derives its name—the local form of Lawine—from the fact that the village has been twice destroyed and repeatedly damaged by avalanches. ¾ hr. farther, 3,376 ft. above the sea, stands

Lermoos (Inn: Post. a tolerably comfortable country inn), lying on the slope of a hill overlooking the E. end of the Gaisthal, which separates the two abovementioned mountain ranges. The road to Partenkirch is described in Rte. F. The scenery here assumes a character of unexpected grandeur. The basin, about 3,250 ft. above the sea, extending S. to Biberwier and E. to Ehrwald, was once filled by a lake, in the midst of which rose rocky islets. Above towers boldly the mass of the Zugspitz, bare of verdure from the summit to the base. It was ascended from this side without much difficulty by Messrs. R. Pendlebury, Taylor and Green, in 1870. The brothers Rauch of Lermoos are good guides. In the opposite direction, W. of Lermoos, is the Gartner Wand, often visited by botanists.

The pedestrian may find his way from hence to Zirl in the Innthal by the Gaisthal (Rte. F), or he may rejoin the route described below by traversing a fine pass over the range of the Miemingergebirge, nearly due S. of Lermoos. The path passes the *Sebersee* and another mountain tarn called the *Drachensee*, both well stocked with fish, and descends on the opposite side to Weissland near Obsteig (see below). The lover of nature cannot, however, select any more attractive route than that of the Fern Pass, which is traversed by the high road to Nassereit. This part of the road, as well as that leading from Reutte to Lermoos, are seen to much more advantage by the traveller approaching Lermoos than when taken in the opposite direction.

The road after leaving Lermoos descends the slope to reach *Biberwier*, and almost immediately commences the steep but not long ascent leading to the Fern Pass. The road mounts a wooded ridge that divides two pretty lakes— *Weissensee* and *Mittersee*—and at a higher level looks down into a deep hollow enclosed between rock and pine forest, wherein lies the *Blindsee*, and whence it has no visible outlet. On reaching the plateau, the traveller must not fail to turn round and enjoy the grand and beautiful pictures that are opened towards the E. and NE. After going for some distance nearly at a level, the traveller passes a wayside inn, and beyond it an oratory and an iron monument to Ferdinand I., which mark the summit of the *Fern Pass* (4,063′). The descent commences through a cleft in the limestone rocks; the new road keeps farther east, while the old road on the rt. hand is better for pedestrians. Close under the old castle of Fernstein, on the new road is a new and tolerably good inn, whence a boat will ferry the traveller across a pretty little lake to the ruined castle of Sigmundsburg, under which passes the old road. S. of the lake and of the summit of the pass the two roads reunite, and, amid scenery scarcely less beautiful than that on the opposite slope, descend into the open valley wherein stands

Nassereit (Inn: Post, much improved, as good as any of those on this road), 2,776 ft. above the sea, at the head of the Gurglthal leading to Imst (Rte. F). In the opposite direction, nearly due E., is the way to Innsbruck. The road enters a narrow hollow in the mountain called Rossbach, and then ascends in zigzags till it gains the level of an undulating plateau, a sort of broad terrace, about 3,000 ft. above the sea, at the S. base of the range of the Mundistock, extending thence to the valley of the Inn. At its E. end, between Miemingen and Telfs, this terrace has been levelled as though by the action of water.

At *Obsteig*, where the path from the Drachensee (mentioned above) rejoins our rte., the ridge of the Tschürgant (Rte. A), dividing the valley of the Inn from the Gurglthal, subsides into the plateau, and the mountain is more easy of access from this side than from Imst. The stream flowing past Obsteig soon turns S., and descends through a narrow ravine to Möz in the Innthal. The same course was followed by an ancient road, of which some traces yet remain, guarded by a solitary tower of the ancient castle of Klamm. Near the road E. of Obsteig are the ruins of *Freundsheim*, once belonging to the historic family of Freundsberg.

Mieming is the collective name for a number of hamlets scattered over the plateau. That called Ober Mieming is the post-station, and has a good country inn. The way now descends gradually, in great part under the shade of larches and pines, till it suddenly breaks into the valley of the Inn at *Telfs*, where it joins the main road to Innsbruck described in Rte. A.

Route C.

REUTTE TO LANDECK, BY THE LECH THAL.

Ordinary travellers going from Füssen and Reutte to the upper part of the Inn valley will follow the road last described to Nassereit, and that leading thence to Imst and Landeck (Rte. F). The mountaineer may prefer to traverse one or other of the comparatively high passes over the range dividing the Upper Lech from the Inn, which are approached through the lateral glens on the S. side of the Lech Thal. The char-road through that valley is described in § 41, Rte. E. As to these passes, very little information has reached the Editor beyond that found in Schaubach's well-known work. To reach the opening of each glen, the traveller may avail himself of the post-carriage which runs three times a week from Reutte to Steg.

1. *By the Rothlechthal.* To reach this glen the traveller may keep to the road along the l. bank of the Lech as far as Weissenbach, or may follow a cart-track by the rt. bank till he reaches the opening, just opposite to the last-named village. The entrance is by a singularly wild and savage defile, but after a while the glen opens out, and the traveller finds with surprise that it is divided only by low and gently sloping ridges from the valleys on either hand, so that the hamlets at the upper end, the chief of which is *Brand*, are more easily reached from Büchlbach and Bärwang from the E. or from Namles to SW. than by the path from Weissenbach.

From the upper end of the Rothlechthal a path goes to Nassereit, over the *Dirschentritt Joch* (4,148'), close under the Kamplesspitz, joining the road from the Fern Pass, 2 m. above that village. Another more difficult way leads southward to Tarenz above Imst. From either place the traveller reaches Landeck by the high-road (Rte. F).

2. *By the Naml serthal.* This glen opens into the Lech Thal at Stanzach. It is similar in character to the last, but the village of *Namles* (4,008'), or Namlos, reached after issuing from the narrow defile at its lower end, is more considerable than any of the hamlets of the Rothlechthal. The way to Imst lies nearly due S. from that village, and joins the mule-path through the Pfafflarthal near the summit of the pass next described.

3. *By the Pfafflarthal.* This glen, drained by a torrent called Streinebach, is the most populous of those on the S. side of the Lech Thal. It opens a short way above Elmen. A mule-track leads all the way from that village to Imst. About 2 hrs. above the opening is *Bschlaps* (4,397'), the chief village, succeeded by other hamlets, of which the highest is *Pfafflar*, lying at the N. base of the *Muttekopf* (9,077'). The mule-track is carried eastward, and then winds round the shoulder of that mountain in the descent to Imst, reached in about 8 hrs. from Elmen. The mountaineer might probably take the summit on his way; but if bound for Landeck his shortest course is by a hunter's path over the ridge W. of the Muttekopf, descending to Mils in the Oberinnthal. For this route a guide would be indispensable.

4. *By the Gramaisalthal and Zamserjoch.* The glen of Gramais opens into the Lech Thal at Heselgehr below Elbigen Alp. 2 hrs. from the junction, is *Gramais* (4,408'), where the curate receives strangers. Bearing to the l. from the head of the glen (path partly overgrown with *Krummholz*), the top of the Zamserjoch, commanding a fine view, is reached in 2¾ hrs. The descent may be shortened by leaving the path where it turns to W. through a lateral glen, and 3 hrs. or less suffice to reach *Starkenbach*, about half-way between Imst and Landeck. (See Rte. A.)

5. *By the Lendthal.* The Lendthal, also called Madauthal, lies opposite the hamlet of Lend, between Holzgau and Elbigen Alp. About 2 hrs. from the opening the track reaches *Madau*. Here three torrents, from as many different branches of the glen, meet together. The middle branch, called Parseyerthal, leads to the *Parseyerspitz*, which must command a fine view of the neighbouring valleys. The rt. hand or SW. branch, called *Alpenschonerthal*, is traversed by a path which after crossing the ridge at its head descends to Schnaun, on the Arlberg road, about 12 m. above Landeck. In descending from the pass the traveller passes through the remark-

able cleft called the *Schnauner Klamm.* The direct way from Madau to Landeck lies through the l. hand or SE. branch of the Lendthal, called *Reththal*, at the head of which is the *Matrioljoch*. The descent is through the Matriolthal, drained by the Letzerbach, and the path before reaching the high road near Landeck passes the remarkable waterfall mentioned in Rte. A.

Further information as to all the passes above mentioned is much desired. In fine weather they must command very favourable views of the snowy range of the Œtzthal Alps.

A traveller wishing to see the head of the Lech Thal, and not objecting to a detour, may go by Lechleiten and Stuben, or else by Steg and the Kaiserjoch. See § 41, Rte. A.

Route D.

MUNICH TO INNSBRUCK, BY PARTENKIRCH.

	Post miles	Eng. miles
Starnberg . . .	3½	16¼
Weilheim . . .	3½	16¼
Murnau . . .	2½	11½
Partenkirch . . .	3¼	15
Mittenwald . . .	2¼	10¼
Seefeld . . .	2½	11¾
Zirl	2	9½
Innsbruck . . .	1¾	8½
	21¼	99

Railway to Starnberg; post-road thence to Innsbruck; Bavarian miles as far as Mittenwald, Austrian miles from that place to Innsbruck. For the lake steamer, see below.

Since the opening of the rly. to Starnberg, and the establishment of a steamer on the lake, the Starnberger See has become the favourite outlet of Munich, and many villas belonging to the gentry and citizens stud its shores. Trains run three or four times a day in one hour from the capital to the village of

Starnberg (Inns: Pellet; Tutzinger Hof; Post), and in connection with the first morning train a diligence or post-omnibus runs, or did run, to Mittenwald. It travels slowly (53 m. in more than 11 hrs.), and, except from the coupé, the traveller sees nothing of the beautiful scenery through which he passes. In this part of Bavaria the most agreeable mode of travelling is to post. If not overburdened with luggage, the traveller will be provided at each station with a neat and comfortable open carriage, and the open spaces of green pasture broken by clumps of pine forest will often give him the impression that he is driving through an English park. Such is the general character of the scenery between Starnberg and the town of

Weilheim (Inns: Post, good; Braüwastl). Besides the diligence above mentioned, an omnibus runs three times daily to the rly. station. About 7 m. to the SW. is the isolated hill called *Hohe-Peissenberg* (3,231′), rising about 1,000 ft. above the plain, and commanding a view of surprising extent considering its very moderate height. The omnibus plying from Weilheim to Peiting (Rte. E) passes by the N. base of the hill, whence the summit is reached in ¾ hr. Beyond Weilheim the road to Innsbruck is carried due S., and after passing between two small lakes—Staffelsee and Riegsee—reaches

Murnau (Inns: Post, good; Gries braü), a clean thriving market-town, which has been almost destroyed by fire three different times within thirty years.

Instead of following the post road

many travellers prefer to reach Murnau by the steamer on the *Lake of Starnberg* (1,947′), also called *Würm See*, from the name of the petty stream which drains it. Though one of the largest lakes in Germany—15 m. long, 3 to 4 m. in width— it is fed by none but a few trifling rivulets, as the drainage of the surrounding district is carried off either by the Isar to the E. or the Amper on the W. side. The lake is enclosed by low hills, on whose slopes are many villas surrounded by gardens or parks. The scenery would be tame were it not for the bold forms of the Bavarian and Tyrolese Alps forming a noble background to the picture. A small steamer plies twice a day from Starnberg, making the circuit of the lake. The traveller bound for Murnau will land at *Seeshaupt* (good dinners, with fine lake-fish, at the inn near the landing-place), the village at the S. end of the lake. Omnibuses run from hence to Murnau and Partenkirch, to Benedictbaiern and Kochel (Rte. G), and to Tölz (for Tegernsee). It is prudent to secure places in these omnibuses on board the steamer.

The road from Murnau to Partenkirch is carried for the most part along the l. bank of the Loisach. For several miles a tract of morass lies on the W. side of the road, which fairly enters among the mountains at *Eschenlohe* (a good roadside inn). On a projecting rock are some remains of the ancient castle of the Counts of Eschenlohe, whose line became extinct at the beginning of the 13th century. Here the valley of the Loisach is contracted between the base of the *Krottenkopf* (6,882′) to the E., and the *Ettaler Mandl* (5,798′). At *Oberau* (very good inn) the valley, which seems the filled-up bed of an ancient lake, opens out, and the road from Ammergau (Rte. E) joins our rte. About 3 m. farther the road crosses the Loisach near the ruined castle of *Werdenfels*, and soon leaving the river on the rt. hand reaches

Partenkirch (Inns: Post, good, civil people; Bär; Brauhaus; Stern), an ancient town beautifully situated, 2,369 ft. above the sea, in the neighbourhood of the highest peaks of the Bavarian Alps. It is frequented by visitors in summer, and the inns are sometimes full. Fair quarters and very reasonable prices are to be had at the inn 'Zum Husaren,' in the adjoining village of Garmisch. The sportsman or naturalist who would halt here to explore the neighbourhood may find clean and comfortable lodgings in private houses. The stranger who would learn something of the national manners and dialect should visit the Gastzimmer, or common taproom, of a large inn in the evening, especially on a holiday. Music, singing, and the characteristic national dances are kept up till a rather late hour. The woollen jackets, called Juppen, are now often bought by strangers, but seldom found ready-made. They are very warm and extremely cheap (costing about 8s.). About a mile from the little town is *Kainzenbad*, or Kanitzerbad, a bathing establishment frequented for its slightly sulphurous spring. There are many spots in the neighbourhood commanding fine views; of those near at hand the St. Antonskapelle is probably best deserving a visit.

The most interesting excursions from Partenkirch are those which penetrate the recesses of the Wettersteingebirge, especially those to the Rainthal and Höllenthal, noticed in Rte. F.

A short afternoon stroll may be made to the Partnachklamm. The *Partenach* torrent issues from the Rainthal, being fed by the glaciers at its head, and joins the Loisach between Partenkirch and Garmisch. A country road is carried some way along the stream; following this, the stranger finds a fingerpost directing him to the 'Klammbrücke.' This is a wooden bridge thrown over the chasm cut by the torrent at 222 ft. above its level, and commands a remarkable view. Ascending for 10 min. from the bridge and bearing to the l., the traveller reaches a woodman's cottage, 'Auf dem Graseck,' where refreshments may be had. So far the walk (going and returning) will take about 2 hrs. Those who have no objection to double that distance may go from the Graseck to a point called Ecklauer, and return to Partenkirch by the Kainzerbad.

The *Krottenkopf* (6,882′) is often ascended for the sake of its extensive panoramic view, which embraces six or seven lakes and a wide range of Alpine summits. The way is by the Esterbergalp, where Alpine fare and hay for nightquarters are found.

The best guide at Partenkirch is Reindl, surnamed 'Der Spadill.' He is an intelligent man, and has drawn up a short account of the various mountain excursions in the neighbourhood of Partenkirch. Joseph Koser of Garmisch and his brother have been recommended. The first is a good guide, but exacting and tricky in money matters.

Between Partenkirch and Mittenwald the high road passes from the valley of the Loisach to that of the Isar through one of those troughs running E. and W. which are characteristic of the orography of this district. The road ascends from Partenkirch to the watershed, but on the E. side the descent is very slight. On entering the valley of the Isar, which shows a broad expanse of green meadows, the road bends to SE., and soon reaches

Mittenwald (Inn: Post, pretty good, not equal to those at Partenkirch or Walchensee), a large village on the l. bank of the Isar, 3,010 ft. above the sea, lying in the deep cleft that divides the E. end of the Wetterstein range from the W. extremity of the Karwändlgebirge, further noticed in § 43. The highest summit of this range the *Karwändlspitz* (8,259′)—rises very boldly at the E. end of the range, and is not so well seen from Mittenwald as from more distant points. This mountain village derives considerable profits from the fabrication of violins, guitars, and other stringed instruments, which are chiefly exported to England and America.

S. of Mittenwald the road follows the nearly level course of the Isar, crossing that stream before it receives from SW. the torrent from the Leutaschthal (Rte. F). The frontier between Bavaria and Tyrol follows the crest of the mountains on either side, but extends a short way southward through the valley to a point where it is contracted to a narrow defile, called in the middle ages *Porta Claudia*, from a fortress built by Claudia de' Medici, widow of Ferdinand V., Count of Tyrol. In 1805 a small garrison of 600 Austrians repulsed a first attack made on this strong position by 13,000 French troops under Ney, but finally capitulated, whereupon the fort was razed to the ground. On crossing the frontier the road enters the village of *Scharnitz* (3,138′), which gives its name to the defile. Extra horses are taken here for the ascent to Seefeld. Near the village the Isar is formed by the junction of the torrents flowing westward from three parallel glens—Karwändlthal, Hinterauthal, and Gleirscherthal, further noticed in § 43, Rte. G. Through the last the mountaineer may find a direct but laborious way to Innsbruck.

At Scharnitz the road quits the banks of the Isar, and begins to ascend along the Raabach, a stream which flows towards NE. from the plateau of Seefeld. From several points, especially from the ruined castle of Schlossberg, there are fine views of the Karwändl range to NE., and the Hohe Mundi to W.

Seefeld (Inn: Post) stands at 3,900 ft. on the flattened summit of a broad col dividing the range of the Mundi Stock, or Miemingergebirge, from that of the Solstein. The bituminous shales of the neighbourhood yield a considerable quantity of petroleum. A country road leads hence nearly due W. to Telfs, about 9 m. distant. The road to Innsbruck now turns SSE. and begins to descend towards the valley of the Inn. the declivity on this side being much longer and steeper than towards Scharnitz. After numerous zigzags that afford pleasing views of the broad valley below, the traveller passes beneath the remaining tower of Maximilian's castle of Fragenstein, and reaches

Zirl. For the road thence to Innsbruck, see Rte. A.

Route E.

MUNICH TO PARTENKIRCH, BY AMMERGAU.

	Bavar. miles	Eng. miles
Inning	4¼	19½
Bayerdiessen	2¼	11¼
Weilheim	2	9¼
Peiting	3¼	15
Ober Ammergau	4	18½
Partenkirch	2	9¼
	18	83

The chief stream flowing northward from the Alps through the plain of Bavaria between the Lech and the Isar is that which drains the *Ammersee*. On issuing from the lake the river is called *Amper*, but in its upper course from the mountains to the lake it is known as the *Ammer*, and the valley as the *Ammerthal*, except its uppermost extremity, locally called Graswang Thal. Ammergau, the chief place in the valley, has of late become well-known in England from some interesting accounts of the singular dramatic representations of the Passion and Crucifixion which are here periodically repeated at decennial intervals. Except at such periods the valley is rarely visited by strangers, although it was in the middle ages the common route for Venetian merchants travelling to Augsburg, the portion between Peiting and Ettal being, in truth, on the direct line from that city to Innsbruck.

The Ammersee is reached directly by the post-road from Munich, through Pfaffenhofen (post-station) to *Inni g,* close to the N. end of the lake; or the same place may be reached by a cross-road from the Maisach station on the line from Munich to Augsburg—distance about 12 m.

A better way than either of these to reach Inning is to take a country road from Starnberg (reached by rly. from Munich), pass the castle of Seefeld at the NE. end of the *Pilsensee*, and round the S. and W. shores of the silent *Wörthsee* girdled by forest. The scenery of the Ammersee is pleasing, but not equal to that of the Lake of Starnberg. There is a tolerable country road along the W. shore to

Bayerdiessen (more commonly called Diessen), a post-station, whence runs a good road to Fischen, Pähl, and Weilheim. The traveller who has preferred to traverse the lake in a boat may land at Fischen, and there join the last-named road. On a height above the E. shore of the lake he will observe the convent and ruins of the castle of *Andechs*, which in the 11th and 12th centuries was the seat of the powerful family who bore that name.

At Weilheim (Rte. D) the road from Starnberg to Murnau approaches close to the Ammer, but the traveller intending to reach the upper part of that river will either follow the post-road to *Peiting* (Inn is, or used to be, excellent), leaving the Hohe Peissenberg on his l. hand between him and the river, or else will take a country road by the rt. bank, but mostly at a distance from the stream, passing through Böbing, and rejoining the post-road from Peiting at *Rottenbach*. Before reaching that place, the valley of the Ammer, which for 8 or 9 m. had followed an upward course nearly due W., makes a sharp turn, and thence to Ettal the general direction is always SSE. The hamlet of Echelsbach contained in former times a very large inn, or caravanserai, whereat travellers from the south, choosing for greater safety to move in a numerous company, halted on the way to Augsburg. Beyond this the road diverges eastward from the Ammer, passes

Bayersoyen and *Saulgrub*, rejoining the stream, and crossing to the l. bank before it reaches

Unter Ammergau (2,659′). For some distance below the village the valley has been enclosed between mountains of considerable height, though less bold in form than those lying nearer to its head. To W. is the range of the *Trauchberg*, culminating in the *Hohe Bleichberg* (5,372′), while on the opposite side the highest summit is the *Aufacker* (5,132′). It is said that good country inns are found here as well as at

Ober Ammergau (2,762′), 3 m. farther up the valley. This, the principal village of the valley, stands close to the base of the Kofel (whence the Roman name of the station, Arces Coveliсæ?), the E. extremity of a high range extending due W., parallel to the Trauchberg, nearly to Hohenschwangau (Rte. B). Its highest peak is the *Klammspitz* (6,840′). The *Ettaler Mandl* (5,798′), on the E. side of the valley, is exactly in the prolongation of the axis of this range. The population here are largely employed in the production of carved ware (Germ. Schnitzwaaren) in wood and stag's-horn, and are also painters on glass. It has been reasonably conjectured that the education thus indirectly conveyed has contributed to the singular artistic merit of the representation of the Passion, for which this village is now widely known, and which led hither an extraordinary concourse of strangers on the last occasion when it was publicly given.

The unanimous report of eye-witnesses declared the performance to have been relieved from what would otherwise have been incongruous and offensive by the genuine faith and earnestness of the performers, while its positive merit in an artistic sense surpassed all expectation. In the year fixed for the purpose, the representation is repeated in the open air on every Sunday and Monday during the height of summer.

Ober Ammergau is but 2 hrs. from Oberau, on the road from Murnau to Partenkirch. The way lies through a deep depression on the E. side of the Ettaler Mandl. Just half-way is the ancient monastery of *Ettal*, founded by Louis of Bavaria in 1330, and suppressed in 1803. It is now a private residence. It contains several paintings of some merit and a celebrated marble statue of the Madonna brought from Italy by the founder.

The descent from Ettal to Oberau is rather rapid, as the level of the Loisach is lower than that of the Ammer.

The pedestrian who would follow the latter stream to its source must turn due W. from Ettal and ascend the *Graswang Thal* with the range of the Klammspitz (6,840′) to the N., and a more broken and irregular mass whose highest summit is the *Kreuzspitz* (7,156′) to the S. Near the hamlet of Linder the glen divides. One branch, traversed by the Säger torrent, mounts due W., and leads to Hohenschwangau. The other stream, called Linder, descends from SW. The pass at its head leads to the Plan See (Rte. B) through the glen of the Ammerwald. Having reached the E. end of the Plan See, the traveller may turn eastward by a track leading past the Eibsee to Garmisch and Partenkirch (Rte. F), or he may descend to Reutte by the Stuibenfall, or follow the banks of the lake to Heiterwang by the path noticed in Rte. B.

Route F.

PARTENKIRCH TO IMST IN THE INNTHAL.

The traveller going from Munich to Switzerland by the Engadine, or to Lombardy by the Stelvio Pass, must aim at Imst in the Oberinnthal as a place that unavoidably lies in his way. It is true that the shortest way, reckoning by time, is by the rly. to Innsbruck, and thence by the high road described in Rte. A.; but a more direct, and on the whole a more interesting route is that by the Lake of Starnberg to Murnau and Partenkirch, and thence by Lermoos and Nassereit to Imst. The first part has been given in Rte. D., and the portion from Lermoos to Nassereit in Rte. B. The remainder of the way is now to be described. The mountaineer who observes that a straight line from Partenkirch to Imst runs through the centre of the Wettersteingebirge, the highest group of the Bavarian Alps, will, however, be loth to follow a carriage-road, and may be tempted to prefer some one of the mountain tracks noticed below.

1. By the *Loisachthal*. A carriage-road; 3½ Bavarian m. to Lermoos; 4 Austrian m. thence to Imst.

The Loisach, one of the principal affluents of the Isar, is formed by the confluence of several small streams near Lermoos (Rte. B). It flows thence for several miles in a direction somewhat E. of N., till it is joined by a short stream from the Griesen Pass, and turns due E., but, on approaching Partenkirch, it resumes its course to NNE. A good road is now carried along the valley from Partenkirch to Lermoos, a distance of 16 Eng. m. The village of *Garmisch* (Inn: Zum Husaren, very fair and reasonable) is little more than a mile from Partenkirch; when this is left behind, the Loisachthal shows little trace of man's presence. The scenery is wild, but somewhat wanting in variety, though fine views are gained at intervals of the very steep N. and W. sides of the Zugspitz. A solitary inn stands at the point where the road begins to turn SW., and the track to the *Griessen Pass*, leading to the Plan See and Reutte, is seen due W. Soon after the road crosses the frontier and enters Tyrol. The charge for a one-horse country carriage between Partenkirch and Lermoos is 4 florins. Instead of going to Lermoos, the traveller may save a little time by avoiding that village, and following the cross-road by *Ehrwald* (3.296'), joining the high road at Biberwier. The charming scenery of the road by the Fern Pass from Lermoos to Nassereit has been noticed in Rte. B.

From Nassereit to Imst the way lies due SW. down the *Gurglthal*. The slopes of the Tschürgant to the rt. of the road are somewhat monotonous, but on reaching *Tarenz* (2,776'), where the track from the Lech Thal by Bschlaps and Pfafflar (Rte. C) joins the road, a fine view of the surrounding mountains opens out. It is rather more than 9 Eng. m. from Nassereit to Imst (described in Rte. A).

2. *By the Eibsee*. The moderate walker, who is not disposed for the comparatively laborious excursions next pointed out, may take on the way to Lermoos the mountain lake called *Eibsee* (3,211'), lying very near the N. base of the Zugspitz. It contains a number of picturesque wooded islands, and has no apparent outlet. There is a char-road from Garmisch to *Ober Grainau* (2,541'), where a guide is easily found. The charge for a guide going to the lake (1 hr.), and thence as far as the road to Lermoos at the frontier (2 hrs.), is only 48 kreutzers. Two families of fishermen (said to be descended from gipsies) live by the lake, and endeavour to gain some profit from visitors. M. Bädeker advises travellers to keep clear of them, and especially to avoid hiring any of them as guides. The huge mass of the Zugspitz, rising 6,500 ft. above the lake, falls rapidly on the NW. side to the comparatively low ridge of the *Thörleswand* (5,227'), over which a frequented track is carried to Eerwald.

3. *By the Höllenthal.* The mass of the Wetterstnngebirge consists essentially of two parallel ridges welded together at their W. extremity, where they form a single mass culminating in the Zugspitz and cut away very abruptly on the W. side. A third ridge, whose chief summit is the *Waxenstein* (7,409'), diverges from the N. side of the Zugspitz towards ENE. The Alpine glen dividing the two first-mentioned ridges is the Rainthal, that lying between the central ridge and the Waxenstein is the Höllenthal. Both glens terminate in rather considerable glaciers, and one or both of them should be visited by the mountaineer who would become thoroughly acquainted with this group. Good guides are the Rauch brothers of Lermoos and Reindl of Partenkirch. The way to the *Höllenthal* is by a path leading from Garmisch to Hammershach, a hamlet having the same name as the torrent issuing from the glen. A path leading to an abandoned lead-mine conducts the stranger to a very remarkable cleft called the Höllenthal Klamm, which is spanned by a wooden bridge. From this point the mountaineer may ascend the *Alpspitz* (8,648'), or the somewhat lower and more westerly peak of the *Höllenthorspitz,* returning to Garmisch or descending on the opposite side into the Rainthal. From the Klamm the traveller may also reach the glacier (*Höllenthalferner*) at the head of the glen. The scenery is extremely bold, and of the wildest character. The most direct way to reach Ehrwald and Lermoos from the Höllenthal is to cross the ridge of the Waxenstein, and descend on the N. side to the Eibsee. The brothers Rauch of Lermoos are acquainted with the way. The ordinary excursion is by a pass called *Kreuzjoch,* leading to the Rainthalhof, mentioned below. This easy pass commands fine views of the near peaks and precipices, as well as the distant ranges of the surrounding Alps.

4. *By the Rainthal.* Although the scenery of the Höllenthal is very wild and grand, it is surpassed by that of the *Rainthal,* or Reinthal, already mentioned as the glen dividing the two principal ridges of the Wetterstein group. Rough accommodation for the night may be found at the *Rainthalhof* (3,092'), 2 hrs. from Partenkirch, but provisions should be taken, as nothing beyond chalet fare will be found there. The best arrangement for the mountaineer is to start early from Partenkirch, and, after leaving his superfluous baggage at Rainthalhof, to ascend by the N. side of the *Dreythorspitz* (8,463') to the Schachenalp, and thence reach the *Teufelsgesäss* (7,008'), commanding one of the grandest views in this district It is possible to pass the night at the Schachenalp, but more advisable to return to Rainthalhof. Starting early on the following day, the traveller should ascend the Rainthal. After 1 hr. he issues from the forest into the savage defile extending to the *Plattacher Ferner,* the glacier that lies in the background during the ascent. For some way the track lies in the bottom of the cleft, where the stream is at one spot bridged over by a mass of avalanche snow. The most striking points of view is from a point called Auf der Platte, where the glacier is seen in the centre of a vast amphitheatre of bare rocks rising in successive tiers to the summits of the *Wetterschroffen,* or *Plattspitz* (8,862'), the *Schneefernerkopf* (9,415'), the Zugspitz (9,716'), and the *Brunnthalspitz.* A hunter's pass called *Gatterl* (6,638') crosses the southern side of this great cirque. The pass lies between the Wetterschroffen and the fine peak of the *Wanner* (8,997'). The ascent commences near the point where one of the sources of the Partenach bursts out from a cleft in the limestone rock. The descent is by the steep ledges of the Wetterschroffen, till it reaches the path from Mittenwald to Ehrwald (further noticed below), close to the Pestkapelle.

The traveller, who makes the excursion to the Rainthal intending to return to Partenkirch, may visit the head of the glen on the first day, sleeping at Rainthalhof, and on the following day make a circuit by the Stuibensee, Kreuzjoch, Höllenthal, and Garmisch. An additional day may be allowed for a visit to the Teufelsgesäss. Another arrangement is to go on the first day by the Höllenthal, Kreuzjoch, and Stuibensee, to Rainthalhof; on the second to visit the head of the Rainthal, cross the pass Auf der Leiter, then turning eastward to the village of Leutasch, and to return on the third day, either by the road through Mittenwald, or by the Ferkenbach.

The noble peak of the *Zugspitz* (9,716′) is best reached from the Knorrhütte at the upper end of the Rainthal, 6 to 7 hrs.' walk from Partenkirch. It has a double summit—the western being by a little the higher—connected by an impassable ridge of jagged teeth of rock. To reach the W. peak, marked by a large cross, the mountaineer goes on the first day from Partenkirch to the Knorrhütte; starting early next day, he ascends by the glacier and a rocky aréte in 2½ hrs. to the top. The guides used to ask only 4 flor. a day and their food. The view is said to include the Finsteraarhorn and Mont Blanc (?), and certainly the whole Tauern range and the Grisons Alps.

5. *By the Leutasch Thal and Gaisthal.* A more circuitous route from Partenkirch to Ehrwald than any of those above described is by the depression which separates the Wettersteingebirge from the parallel range of the Miemingergebirge. The watershed between the Loisach and the Isar lies about the middle of this depression; and the two torrents, one flowing W. to Ehrwald, the other E. towards the village of *Leutasch*, both bear the name Gaisbach, and the entire valley is called *Gaisthal*. A little above Leutasch the eastern Gaisthal opens out and bends to ENE., assuming the name *Leutaschthal*, or Luetaschthal, and joining the valley of the Isar just above Mittenwald. Although a circuitous way for the traveller starting from Partenkirch, it is the most direct route to the head of the Loisachthal for a traveller who has reached Mittenwald from Munich by the Walchensee (Rte. G).

Going from Partenkirch, the traveller may take the high road as far as Mittenwald, crossing the Tyrolese frontier, and entering the Leutaschthal within 1 m. of that village; or he may reach the lower part of the same valley by a pass lying at the head of the glen of the *Ferkenbach*, a torrent that joins the Partenach from the E. about 1½ hr. from Partenkirch. The Leutasch is an open valley with several scattered hamlets, the chief of which lies at 3,753 ft. above the sea. The abrupt termination of the range of the Hohe Mundi might here lead a stranger to imagine the valley to be a prolongation of that of the Inn, and in truth a very slight ascent, followed by a comparatively long descent towards the SW., will take him to Telfs in the Innthal—only 1½ hr. from the highest hamlet in the valley. Due W. the Gaisthal opens as a narrow cleft between the high ridges on either side. For the greater part of the way the path lies through forest, without any trace of human presence. On the watershed forming the pass—called *der Geissel*—is the Pestkapelle (4,258′?), an oratory erected in commemoration of the plague of 1646, where the few mountaineers who traverse the pass Auf der Leiter from the head of the Rainthal join the present route. A gentle descent leads to Ehrwald, whence the high road may be joined either at Lermoos or Biberwier. The traveller who does not desire to follow the road over the Fern Pass may turn aside to the l. rather more than 1 hr. above Ehrwald, and reach Nassereit by the Drachensee. (See Rte. B.) The distance from Mittenwald to Lermoos, or Biberwier, is 7 hrs., steady walking.

N. T.   D

Route G.

MUNICH TO INNSBRUCK, BY THE WALCHEN-SEE.

	Post miles	Eng. miles
Baierbrunn	2¼	10¼
Wolfratshausen	1¾	8
Königsdorf	1¾	8
Benedictbaiern	1¾	8
Walchensee	2½	11½
Mittenwald	2½	11½
Innsbruck (Rte. D)	6¼	29½
	18¾	87

Post-road without diligence. Although post-horses are still found at the above-mentioned stations on this, the shortest road from Munich to Innsbruck, it is a shorter and more agreeable course to take the rly. to Starnberg, and the steamer thence to Seeshaupt (Rte. D). From Seeshaupt a diligence starts every day after the arrival of the first steamer for Benedictbaiern (distant 10 or 11m.), and Kochel; and an omnibus (fare 45 kr.) also plies to *Schlehdorf*, at the NW. corner of the Kochelsee. A ferry across the lake (fare 9 kr.) will land the pedestrian on the high road about 6 m. from Walchensee.

Benedictbaiern (Inn: Post) was famous for a Benedictine monastery which after an existence of more than 1,000 years was suppressed in 1803. The building served for several years as the factory for Frauenhofer's then unequalled optical instruments. Near this is a mineral spring—Adelheidsquelle—which has risen into repute owing to the presence of iodine and bromine in its water. On leaving Benedictbaiern the road approaches the base of the bold ridge of the *Benedictenwand* (5,921'), commanding one of the most extensive panoramic views on the N. side of the Alps. The summit is reached without difficulty from

Kochel (Inns: Bad Kochel, kept by Dessauer; and another kept by Fink). Here the road passes between the W. end of the Benedictenwand and the pretty little *Kochelsee* (1,963'). At the opposite side of the green basin rises the *Heim-garten* (5,881'). At the SE. end of the lake the road climbs the rather steep barrier of the Kesselberg, separating this from the *Walchen See*. There are two waterfalls near the road; the path to one of them is pointed out by a finger-post. The traveller who has recently left the central range of the Tyrol will not think it worth while to turn aside for these comparatively trifling cascades.

The *Walchen See* is an extremely picturesque mountain-lake enclosed between bold rocks and dark woods, except at the E. end, whence the stream of the Jachen flows to join the Isar. Its level is 2,630 ft. above the sea, and its form that of a triangle with sides about 5 m. in length. From the hamlet of *Urfeld*, where the road reaches the N. shore of the lake, a country road runs eastward to the Jachenauthal (§ 43, Rte. E), and a horse-path, constructed by the late King Max, leads in the opposite direction to the summit of the *Herzogstand* (5,763'), commanding a view nearly as extensive as that from the Benedictenwand. A ferry-boat plies from Urfeld to Walchensee—fare 12 kreutzers. The road to Mittenwald runs along the W. bank to

Walchensee, a small village with a good inn (Post). The *saibling* (Salmo salvelinus) of this lake is much esteemed, and therefore dear. They are kept in tanks and fed, it is said, on raw meat. A good view of the lake is gained from rising ground behind the inn. Quitting the lake shore, the road crosses the Katzenkopf, which projects as a promontory into the lake, and soon enters the wooded glen through which the Obernacherbach descends from the E. side of the Krottenkopf (Rte. D). The torrent forms a waterfall to rt. of the road. Turning due S. over green meadows the traveller gains a fine view of the Wettersteingebirge and the Karwändl range, and reaches the village of *Wallgau*. If bound for Partenkirch he may turn aside at *Krün* (2,804'), the next village. The high road follows the course of the

Isar to *Mittenwald*. For the road thence to Innsbruck, see Rte. D.

For the traveller who keeps to the carriage road this route is on the whole more interesting than that by Partenkirch.

SECTION 43.

KREUTH DISTRICT.

The boundaries of the district described in the present Section are easily defined. To the W. it is divided from that treated in the last Section by the course of the Isar, and the low pass of Seefeld from Scharnitz to the Inn. The valley of the Inn from Zirl to Kufstein, and the course of the same river when it turns northward from Kufstein to Rosenheim, fix the S. and E. limits. The irregular mass of mountains—broken into many separate ridges—that extends from Seefeld to Kufstein for a distance of about 50 m., lies altogether in the broad zone of triassic and jurassic rocks that form the northern girdle of the Alpine chain. The highest summits lie in the ridges N. of Innsbruck. First comes the *Lavatscherspitze* (9,081′), then the *Birkkorspitze* (8,978′), the *Edkarspitze* (8,911′), &c. The height of the *Solstein* has been much exaggerated. The highest point, sometimes called Kleiner Solstein, reaches only 8,649 ft., and the western peak only 8,333 ft. But few of the summits lying farther E. between Tegernsee and Kufstein surpass 6,000 ft.

For those strangers who are content to forego the grander and sterner beauties of Alpine scenery, and can content themselves with the most varied combinations of rock and pine forest, brawling torrent, dark lake, and green alp, this district offers unusual attractions. On its northern border good accommodation and cultivated society are found on the shores of the pretty Tegern See. A few miles farther S. the sportsman, the naturalist, and the invalid seeking the gentle stimulus of mountain air, find attractive quarters amid the bolder scenery that surrounds Wildbad Kreuth: but many a mountaineer will prefer to either of these frequented spots the excellent country inns found on the shores of the Achensee, one of the wildest and most striking of the mountain lakes of Tyrol.

The numerous minor glens of this district, scarcely important enough to be enumerated in a guide-book, will yet furnish scope for very many pleasant excursions to the stranger who may devote some weeks to exploring its recesses.

Route A.

MUNICH TO INNSBRUCK, BY WILDBAD KREUTH.

	Post miles	Eng. miles
Holzkirchen . .	4½	20¾
Tegernsee . .	2¾	12¾
Wildbad Kreuth .	1½	7
Achenkirch . .	3¼	15
Schwaz . . .	3½	16½
Volders . . .	2	9½
Innsbruck . .	2	9½
	19½	91

Railway to Holzkirchen, post road thence to Innsbruck. Travellers going from Innsbruck take the rly. to Jenbach, where the road to Kreuth leaves the Inn valley. Carriages for Kreuth are found there (beim Neuwirth). The trains from Munich—all slow—take 1½ hr. to reach

Holzkirchen (Inn: Post), whence diligences and omnibuses ply twice a day to Tegernsee and Kreuth. Three times weekly, during the summer, a diligence runs all the way to Jenbach in the Innthal, while on other days travellers who rely on public conveyances must sleep at Kreuth, and travel thence by omnibus, starting early, and reaching Jenbach about 2 P.M. These vehicles may be recommended to those who go abroad to seek discomfort, and do not care to see the country through which they pass.

After leaving Holzkirchen the outer ranges of the Alps gradually rise in the horizon until the traveller fairly enters the hills at Dürrenbach, and immediately after reaches the N. end of the *Tegern See* (2,410′), a pretty sheet of water about 4½ m. long, enclosed between wooded hills that rise about 2,000 ft. above its surface. The road runs along the E. shore to the village of

Tegernsee (Inns: Post, good, close to the lake; Beim Guggemos, clean and reasonable; when these are full—a frequent occurrence—the traveller will be well treated in the adjoining hamlet of Rothach at Scheurer's Inn). This place has for the last half-century been a frequent resort of members of the royal family of Bavaria, and that circumstance, joined to its natural advantages, has caused a constant influx of visitors during the summer season. Lodgings are found here and in the neighbouring villages. The royal Schloss occupies a portion of the ancient Benedictine Abbey, while another portion of the same extensive pile is used as a brewery, said to produce excellent beer. The favourite excursions are to a low hill behind the village called Paraplui, and (by boat) to *Kaltenbrunn* at the NW. corner of the lake, where parties dine at a café and restaurant on rising ground overlooking the lake. Those who seek wide views ascend the *Neureuth* (4,443′), NE. of the village, or the *Fockenstein* (4,368′), on the opposite side of the lake, or the higher summit of the *Hirschberg* (5,619′), SW. of Egern. A longer, but more interesting, excursion is the ascent of the *Wallberg*, or, better still, of the *Risskogl* (6,042′), at the S. end of the same range. From the summit the mountaineer may descend through the Langenauthal to Kreuth, and thence return by road, or else bear eastward till he falls into a track that will lead him to Tegernsee through the glen of the Rothach. For the excursion to the Schlier See, see Rte. C.

The road from Tegernsee through the glen of the *Weissach* lies throughout amid very pleasing scenery. A finger-post points the way to some considerable marble-quarries, and near to these, about ¼ hr. from the road, is the fall of the *Lohbach* torrent descending from the Hirschberg. The stream is slender, but the position very picturesque. At the village, which is distinguished as *Dorf Kreuth* (2,630′), are one or more second-rate country inns, whereat strangers not finding room at the Baths are used to lodge. The glen of the Weissach is here contracted, and the scenery assumes a somewhat more Alpine character when, 2 m. beyond the village, the road reaches

Wildbad Kreuth (2,722′). This is a large institution, mainly intended for the reception of invalids. It belongs to Prince Charles of Bavaria, and is managed by a Hausmeister, the medical direction being under the management of Dr. Stephan, a gentleman of considerable professional skill, who is married to an English lady. 'During the season, which extends from mid-May to mid-September, visitors should write beforehand to the Hausmeister to secure rooms. The accommodation is good, and the charges, fixed by tariff, are very moderate. The mild sulphur spring is little used, goat's whey and fresh herbs being the chief curative agents employed. Rheumatic and neuralgic patients, and persons with weak chests, are those most benefited by Kreuth.' [W. B.] More than 200 visitors are lodged in the establishment, besides which a certain number of the humbler class are housed gratuitously in an adjoining building.

Among many attractive spots in the neighbourhood of the Baths, may be mentioned the Jägerhaus on the rt. bank of the Weissach near the village of

Kreuth. It is adorned with the heads of bears, lynxes, and other wild animals shot in the surrounding district, and is picturesquely placed at the foot of the *Leonhardstein* (4,744'), which towers boldly above the glen of the Weissach. Coffee, beer, and other refreshments are found there. By a spring 10 min. above the Baths is a spring beside which stands a marble bust of King Maximilian Joseph, the founder of the present establishment.

About ½ m. below the Baths the *Sagenbach* torrent descends from the *Langenauthal* to join the Weissach. It forms a pretty waterfall, which is often visited from the Baths—about ½ hr. distant. Ascending eastward through the same glen a path leads to the Pletzeralp, inhabited by herdsmen in spring and autumn. Above the alp the Sagenbach torrent disappears amid the fissures of the limestone, and the glen is devoid of water until near its head at the Langenauer alp, where the torrent, here called Auerbach, again comes into the light of day. A short walk may also be directed to the Gais alp, about 1,100 ft. above the Baths, where the large flock of goats that supply whey to the establishment are pastured. Visitors usually make a circuit from this point to the Königs alp, lying at about the same height, where they find beer and coffee as well as the usual Alpine fare, and then descend to the high road some way south of the Baths: the whole round being about 2½ hrs. easy walking.

A moderate walker will not be satisfied without ascending some of the higher summits that here surround him on every side. He will in the first place be attracted by the bold ridge of the *Planberg* (Mons Platanus), extending from E. to W. along the frontier of Tyrol, and seemingly walling out the inhabitants of the valley from access to that province. The range includes a number of projecting summits, of which the easiest of access is the *Schildenstein* (5,792'), easily reached in 1½ hr. from the Gais alp. After enjoying the view the stranger may return to the Baths by the Königs alp. A slightly more difficult course is to ascend from the Baths along the torrent called Felsenweissach, descending from a wild rocky amphitheatre, whence a steepish path leads up a narrow gorge called the Wolfschlucht, and gains the centre of the ridge of the Planberg. Here the traveller may turn to W., and follow the ridge to the Schildenstein; or, if a tolerably practised mountaineer, he may choose the opposite direction, keeping eastward along the crest until he reaches the highest summit, called *Halserspitz* (6,718'). In descending he may take a path that strikes the ridge a few hundred yards W. of the summit, and leads to the Lahngarten alp; or else keep on his eastward course till he reaches the Baierbach alp, lying on the ridge that divides the Brandenbergerthal (Rte. C) from the Langenauthal. His course homeward is through the latter glen.

Next in height to the Halserspitz is the *Schinder* (6,565'), lying E. of the Baths at the head of the Langenauthal, and commanding an equally extensive view. The *Risskogl* (6,042'), already mentioned among the excursions from Tegernsee, is nearer to Kreuth. Its summit may be reached from the Pletzer alp (mentioned above) by following a track that mounts NE. to the Scheyrer alp, and thence to the Ableithen alp. The way is then straight up to the ridge, which is followed due E. to the summit. In descending the way may be varied by taking a faintly marked track by the Riss alp and Vorderlochberg, whence the middle part of Langenauthal is easily reached.

The *Ross-stein* (6,099') is a remarkable peak with a cloven summit, rising about due W. from the Baths, between the Weissach and the valley of the Isar. Its precipitous eastern face turned towards Kreuth is scarcely accessible, but the top is reached with little difficulty from the W. and N. sides. The best way is by the glen of the Schwarzenbach, which joins the Weissach about 1 hr. above the Baths. A path on the rt. bank leads through forest, past a timber-sluice, up to Alpine pastures. Thence the ascent is continued up a slope to the l. to the Bach alp, lying in a recess of the mountain. The ridge on the rt. of this basin leads to the base of the final peak. 'You climb up a cleft and round the easternmost of its projecting walls, keeping next on its S. side, and cross a deep gully by a narrow neck of rock with precipitous sides, ascending the opposite side to the summit of the higher of the two topmost masses of the Rossstein.' [W. B.] The view is in some respects superior to those from the other mountains of this neighbourhood, owing to the favourable position of the peak in regard to the lakes and rivers of this part of Bavaria. Many other interesting excursions will occupy the mountaineer who makes Kreuth his halting-place, and the variety and beauty of the scenery will surprise many who look with disregard on mountains of such moderate height. Unless a practised cragsman, and well used to find his way alone, the stranger should not ascend the higher ridges without a guide. It is easy to miss the way amidst the pine woods, and in seeking to extricate himself he may often encounter bits of difficult rock climbing.

The botanist familiar with the Swiss and Savoy flora will be interested by finding near Kreuth many of the peculiar species of the Eastern Alps, such as *Arabis pumila, Cardamine trifolia, Silene alpestris, Astrantia gracilis, Achillea Clavennæ, Crepis Jaquini, Rhododendron chamæcistus* (in the Wolfschlucht), and *Allium victorialis*, besides other interesting plants, e.g. *Tozzia alpina, Epipogium Gmelini, Listera cordata*, and *Corallorhiza innata*.

After passing the Baths the road to Innsbruck ascends gently along the l. bank of the Weissach, passes the Klammbrücke and the Schwarzenbach, and about 2 m. farther leaves on one side a gamekeeper's lodge, or Jägerhaus, where visitors from the Baths sometimes lunch. About 6 m. from Wildbad is the Bavarian customs station of *Glashütte*, so named from abandoned glass-works.

There is here a tolerable country inn, where the traveller may generally find conveyance in a rough carriage. A little farther on is Stuben, where a path ascends westward over the low pass that separates the head of the Weissach from the Isar (see Rte. E). Having thus far followed a course parallel to the ridge of the Planberg, the road has now reached a point where that barrier ceases to close the way into Tyrol. From the N. side the ascent to the *Achen Pass* is insignificant, but the descent on the S. side is comparatively rapid. The Austrian custom-house is no longer in the defile of the Kaiserwache, being removed to the village of *Achenwald* (Inn: Traube). Here the road enters the valley through which the *Ache* flows from the Achensee, and bending gradually from N. to W. joins the Isar (Rte. F). The road ascends the valley, twice crossing and recrossing the Ache, which sometimes runs in the bottom of a deep gorge, sometimes through meadows on a level with the road.

The pedestrian going from Wildbad Kreuth to the Achensee may in great part avoid the high road, and at the same time make an interesting walk, by ascending the ridge of the Planberg— taking the Schildenstein on the way if the day be clear—and descending on the S. side of the ridge. He will fall into the short glen of the Klammbach, and may follow the path on the right side of that stream to *Hohlstatt*, where it crosses the high road. But he will do better to turn off from the Klammbach path at a point where a chalet stands in a clearing of the forest, and cross the low ridge that divides the Klammbach from the *Ampelsbach*. The glen drained by this stream opens into the valley of the Ache at *Leiten*, a village on the high road, whence there is a striking view of the *Guffert* (about 7,300'), here showing as a very bold pyramidal peak. To SE. is the *Unnütz* (6,927'), easily ascended, even by ladies, in 2 hrs. from Scholastica's inn. Due W. is the *Juifen* (7,144'), overlooking the lower course of the Ache.

It would appear that the N. end of the original basin of the Achensee has been filled up by the detritus borne down by torrents. Thus has been formed a rather extensive level plain, marshy in some places from inadequate draining. In the midst, at 3,061 ft. above the sea, is

Achenkirch (Inn: Post, good, but often full in summer). There is another very fair country inn, kept by Kern, on the way from Achenwald to Achenkirch, about ¼ hr. before reaching the latter place, but neither is so attractive a halting-place as those by the beautiful

Achensee. This sheet of dark blue water, enclosed by lofty mountains that rise from the water's edge, is the Tyrolese rival of the Bavarian Königssee, and of the Hallstadtersee in the Salzkammergut. Though very beautiful, it must rank after them in point of scenery, but neither of them offer to the stranger such good accommodation as may be found here. Very near the N. end of the lake is a good country inn (zur Scholastica), 1 hr. from Achenkirch, where boats may be hired for the farther end of the lake. Less than 1 m. farther on is *Seehaus*, a large new inn of considerable pretensions, very ill managed, and therefore not comfortable. The level of the lake is 3,066 ft. above the sea, and its depth is said to be nearly 2,500 ft. The high mountain rising immediately above the W. shore is the *Rubenspitz*. On the opposite side the Unnütz (6,927') towers above the N. end, and is connected by a rugged ridge with the *Heiter Stoll* (6,347'), which on one side looks down upon the lake, and on the other upon the valley of the Inn. The road is carried along the E. shore, in some places supported by piers of masonry, in others carried along a mere notch hewn or blasted in the face of vertical rocks. It is so narrow that it is only at certain spots that two carriages can pass, and the rocks overhang so closely that heavy wagons piled high must be carried over the lake in boats. The sternness of the scene is broken by the little green plain of *Pertisau* on the W. shore, where the streams from the *Falzthurnthal* and the *Gernthal*, after running some miles underground, escape into the lake. At this spot, not easily reached save by boat, are three inns, so much frequented in summer, like that at Seehaus, that a stranger has little chance of accommodation unless he has secured rooms in advance. The Fürstenhaus, by the lake shore, is very clean, but singular and semi-monastic in its arrangements. Rough but clean quarters and fair cookery may be had at the Pfändlerwirth and Karlswirth. At the S. end, where no torrent falls into the lake, is the hamlet of Buchau. Here the traveller approaching the lake from the Innthal finds a boatman to convey him to Pertisau or Seehaus. In spite of its great depth the lake is said to be covered with a solid sheet of ice in winter.

The physical geologist will not fail to remark that the barrier of rock that separates the lake from the Innthal rises scarcely 40 ft. above the water's edge. In a few minutes after leaving the lake the descent begins. To the l., above the road, is seen the village of *Eben*, the burial place of St. Nothburga, whose shrine is a frequent resort of pilgrims. The road descends rapidly about 1,400 ft. through the ravine that leads to *Jenbach* (Inns: Toleranz, new, by the rly. station; Post, in the village, carriages for hire; Neuwirth), a station on the rly. from Kufstein to Innsbruck. The way through the Unterinnthal to that city is described in the next Rte.

Ladies who pass some days at Pertisau may enjoy charming walks through the Falzthurn and Gerenthal glens, which run for miles nearly level amidst grand rock scenery. The moderate mountaineer will find agreeable occupation in climbing the surrounding summits. A pleasant way to the Inn valley is to ascend through the Falzthurnthal to the Stakener Joch, and thence drop through the Stallenthal to Stans in the Innthal, about 3 m. above Jenbach. See Rte. B.

Route B.

MUNICH TO INNSBRUCK, BY KUFSTEIN AND SCHWAZ.

	Post miles	Eng. miles
Holzkirchen . .	4½	20¾
Rosenheim . .	4½	20¾
Kufstein . .	4¾	22
Wörgl . .	2	9½
Brixlegg . .	2	9½
Schwaz . .	2¼	10½
Hall . . .	2	9½
Innsbruck . .	1¾	8¼
	23¾	110¾

Railway, with a change of carriages at Kufstein. From the latter place to Innsbruck the pace by ordinary trains is even slower than that usually imposed on German railways. Many will prefer to hire an open carriage at Kufstein or Wörgl, and so enjoy the fine scenery of the Unterinnthal. Austrian miles and Austrian money are current between Kufstein and Innsbruck.

Between Munich and Holzkirchen the rly. traverses the plain of Bavaria in a direction a little E. of S. and then turns ENE. till it meets the Mangfall. This stream, after flowing northward from the Tegern See (Rte. A), and receiving an affluent from the Schlier See (Rte. C), bends round abruptly to ESE., is joined by another considerable stream (the Leitzach, Rte. D), and enters the Inn above Rosenheim. The rly. is carried along the N. bank to the junction station of

Rosenheim (Inns: Greiderer's; Alte Post; König Otto), a clean looking town, where a portion of the brine from the salt springs of Reichenhall (§ 45, Rte. A) is converted into salt, and supplies baths. There is a refreshment room at the station where the trains for Salzburg meet those directed towards Innsbruck. The latter line is carried nearly due S. along the l. bank of the Inn, in great part alongside of the old post road. On the slope of the hill opposite *Raubling* station is the little walled town of *Neubaiern* (or Neubeuern), overlooked by an old castle. The road fairly enters the mountains at the stat. of

Brannenburg. The picturesque and ancient, but still inhabited, castle passed from the Counts of Preysing to a Pallavicini, and is now a factory. The adjoining village (with a good country inn) is frequented in summer by townspeople seeking change of air. Visitors frequent the Schwarzlackkapelle, ½ hr. distant, and commanding a fine view. A more arduous expedition is the ascent of the *Wendelstein* (5,992′). The chapel on the summit is reached in 4 hrs., the upper part being steep. Near the *Fischbach* stat. are seen the ruins of the ancient castle of Falkenstein with some remains of other buildings of still more remote autiquity. Higher up is the church of Petersberg, founded in 1100, with a convent long since destroyed.

Here the Inn becomes the boundary between Bavaria and Tyrol, the rt. bank belonging to the latter. The entrance into the main valley of N. Tyrol was throughout the middle ages beset by numerous castles serving as much for the purpose of organised plunder as for defence. One of the most curious of these is Auerburg, on a rock above the village of Oberaudorf. A little farther on upon the opposite bank is seen the opening of the *Jenbachthal* through which a road runs from the village of Ebbs to Kössen (§ 44, Rte. E.).

The last Bavarian village on the l. bank is *Kiefersfelden*, where on Sunday evenings in summer the country people often have performances in the style of the miracle-plays of the middle ages. Outside the village is a pretty Gothic chapel in memorial of the departure of King Otho on his ill-omened election to the crown of Greece. At this point, where the Thierseerache (Rte. D) issues from the mountains to enter the Inn, rocks approach close to the river, and the post road and railway traverse a defile (Klause) before reaching the station of

Kufstein (Inns: Post, good; Hirsch). The station is on the l. bank near the bridge over the Inn, and the houses there properly belong to the village of Zell. The frontier fortress of Tyrol stands on the opposite side, but modern detached forts guard both banks of the river. The castle is on a detached rock, approached by a single steep path, and ordinary supplies are hoisted up by cranes. The building has of late years served as a state prison, chiefly for political prisoners from the Italian Tyrol.

For nearly 6 m. beyond Kufstein the rly. keeps to the l. bank. It then crosses the river and reaches the

Wörgl station (Inns: Post, tolerable; Lamm). The village stands a little S. of the junction of the stream from the Brixenthal, through which two important roads reach the Inn valley. That from Salzburg by Lofer keeps to the N. side of the stream (§ 44, Rte. A), while the road from the Pinzgau by Hopfgarten keeps the opposite bank. The *Hohe Salve*, called the Rigi of Tyrol (§ 44, Rte. C), is easily reached in 3 hrs. from Hopfgarten (6 m. from Wörgl), or from Itter, which is nearer to the rly. station.

[The pedestrian wishing to reach the Salzburg road from Kufstein may avoid Wörgl, and go direct from Kufstein to Söll by the glen of the Weissache. The walk is said to abound in picturesque scenery, and the houses are said to offer the best examples of the style of carved wood decoration characteristic of Northern Tyrol. A far more laborious way is to mount from Kufstein through the Kaiserthal, ascend the *Hoch Kaiser* (7,611'), and descend to Scheffau on the high road between Wörgl and St. Johann. See § 44, Rte. A.]

On the N. side of the Inn opposite Wörgl is Mariastein, a very ancient castle, with a chapel visited by pilgrims, lying in a very beautiful wooded glen.

After passing *Kundl*, at the opening of the *Wiltschenau*, a glen descending northward from the *Sonnenjoch* (7,271'),* the rly. passes very near the solitary church of St. Leonhard, built in the 11th century by the Emperor Henry II., but restored or rebuilt in a very peculiar style in the 15th century. The stone carving deserves notice. Opposite the opening of the Brandenbergerthal (Rte. C), the rly. passes through a very deep cutting through the live rock, close to the little ancient mining town of *Rattenberg* (Inn: Lederer Bräu, not good). It owed its former importance to a now exhausted silver mine. The rly. station is at

Brixlegg (Inns: Beim Judenwirth; Herrenhaus), where the copper ore brought from Kitzbühel and other mines is smelted. Here the post road keeps to the rt. bank, while the rly. crosses to the N. side of the Inn. This part of the Inn valley, and the adjoining Brandenbergerthal noticed in Rte. C, have yielded many rare plants. The most successful explorer has been Herr Längst, apothecary at Rattenberg, who discovered a second habitat for the very rare *Carex tetrastachya* in a marshy spot in the woods on the N. slopes of the Innthal.

Travellers bound for the Zillerthal (§ 50, Rte. A) usually leave the rly. at *Jenbach* (Inns: Toleranz, new, near the station; Post; Neuwirth), where the road from Tegernsee (Rte. A) enters the valley. The scenery of the Innthal from hence to Innsbruck is so beautiful that those who travel by railway sacrifice much for the sake of a slight saving of time. The bold forms of the nearer mountains, backed at intervals by the snowy peaks of the central chain, are contrasted with the rich cultivation of the populous valley, where the wooded heights are crowned by slender spires of village churches, or the towers of ancient castles. One of the most remarkable of these is the castle of *Tratzberg*, restored by its owner Count Enzenberg. It is said to have 365 windows. At Stans the

---

* In Northern Tyrol the denomination Joch, which properly means a col, or depression in a ridge, is often applied to an entire mountain.

opening of the *Stallenthal* is seen due W. Two passes near the head of the glen—the *Stanserjoch* and *Stakenerjoch*, lead to Pertisau on the Achensee (Rte. A), and by a third pass over the Kaisergrat, the traveller may reach the head of the Rissthal (Rte. G). A pleasant detour from the high road may here be made by the pedestrian to the priory of Georgenberg, standing on a rock half-way up the Stallenthal. Like other similar foundations in Tyrol it has its legend of a holy knight and a miraculous image of the Madonna. The site is singularly picturesque. In returning, the stranger should follow a path crossing the ridge to S., and enjoy the sudden contrast between the silence of the sequestered glen and the exuberant life of the broad valley that opens before him as he again overlooks the course of the Inn, which is only here and there defaced by the masses of detritus rolled down by mountain torrents. The path leads to the extensive Benedictine monastery of *Viecht*, whither the monks of Georgenberg removed in the last century. It is richly adorned with frescoes, and contains in the library an important collection of ancient books and MSS., chiefly relating to the history of Tyrol. The most interesting objects are, however, a series of sculptures in wood which are amongst the best extant specimens of the art. From Viecht the traveller descends directly to the rly. station on the l. bank of the Inn opposite to

Schwaz (Inn: Post). This thriving town has risen again to prosperity after being reduced to ashes by the Bavarians in the campaign of 1809. It was famous in the middle ages for its mines, which in the course of ninety years are said to have produced more than three and a half millions of marks of silver, and 1,330,000 cwt. of copper. The silver mine has been long since exhausted, but the copper and iron mines are still of some importance. The parish church contains a remarkable monument to Hans Dreyling, director of the mines in the 16th century, the joint work of Colin of Mechlin, and Hans Löffler, better known for their works at Inns bruck (§ 42, Rte. A). The roof of the church is covered with copper plates. The fresco paintings executed about 1514 in the Franciscan church and convent by Rosenthaler, of Nürnberg, deserve notice. The castle of *Freundsberg*, said to have been founded by the Romans, was the cradle of the famous race from which sprang George von Freundsberg, the best general of his day, and one of the first who organised an army into the form which it has retained in modern warfare.

SE. of Schwaz is the *Kellerjoch* (7,633'), the summit of which may be reached in 5 hrs., following the torrent Lahnbach. But this is overlooked by a higher summit to SW., the *Gilfertsberg* (8,201'), which must command a wider view (§ 50, Rte. A). The latter is reached through the *Pillerthal*, which joins the Inn 3 m. above Schwaz. On the opposite side of the Inn is the opening of the *Vomperthal*, further noticed in Rte. G. The next station to Schwaz is *Fritzens*, where the lover of nature may well turn aside from the highway to visit a little woodland district, perfectly sequestered, though near at hand. In this part of the valley the high limestone ridges that form its northern wall recede to a distance of 2 or 3 m. from the river, and the intervening space is occupied by an undulating richly-wooded track called *Gnadenwald*. Here the traveller may wander amongst swelling hills and little dales, here and there coming upon a solitary house or secluded hamlet half lost in the woods, until he issues forth at its W. extremity at the village of Baumkirchen, above Hall.

On the S. side of the main valley the traveller may notice *Wattens*, at the opening of the Wattenserthal; ¾ m. farther is *Volders*. Thence the mountaineer may reach the *Glungetzer* (8,781'), through the *Voldererthal*, and descend to Matrey, on the Brenner road (§ 50, Rte. D). The landlord of the Post at Volders some years ago fitted up the adjoining castle of Aschach for summer visitors.

Since the opening of the rly. few strangers pass the village. It was the birthplace of Anton Reinisch, who, in the struggle against the French in 1797, repeated against the hostile bayonets the heroic act of Arnold von Winkelried. The next station is *Hall* (Inns : Krone; Bär), an ancient town, deriving its name and its importance from the salt mine at the upper end of a ravine (Hallthal) that here opens into the Inn valley. The mine, reached in 3 hrs. from the town, lies 4,818 ft. above the sea, between the *Zunderkopf* (6,428'), and the Speckkorspitz (8,649'). An order for admission and a suitable dress is obtained at the chief office in Hall. The saturated brine (here called *sur*), is brought down in pipes to the town, where the salt is obtained in evaporating pans. The history of this little town exhibits an extraordinary series of public calamities. Thrice during the middle ages the heroism of the inhabitants drove back the Bavarian invader, but was powerless against two visitations of the plague, a destructive inundation in 1518, a devouring flight of locusts in 1560, and an earthquake in 1670, which, commencing with violence, continued at intervals with enfeebled energy for two years. Five or six times the town was destroyed or grievously damaged by fire. Worst of all was the war of 1809, when the dreaded French and hated Bavarians united to carry fire and sword into the heart of Tyrol. Not unmindful of their ancient fame, the people of the valley, under the leadership of the brave and skilful Speckbacher, whose monument is seen in the church, equalled or surpassed the deeds of their fathers. Three times in that year the enemy was driven out, and this important position, commanding one of the chief bridges over the Inn, was wrested from his grasp.

On the S. side of the Inn, rather more than half way between Hall and Innsbruck, is the castle of *Amras*, or **Ambras**, one of the most important in **Tyrol.** The ancient building was enlarged by the Archduke Ferdinand II. of Tyrol, and richly fitted up as a residence for his wife, the beautiful Philippina Welser. The library and collections of pictures, engravings, and ancient armour were celebrated, and the castle became the resort of many learned men. When Amras ceased to be the residence of a court the collections were gradually dispersed, and the precious armour was removed to Vienna in 1796 to save it from French rapacity. Some objects of antiquity and good specimens of local wood carving are still seen in the castle; the architecture is uninteresting, and the chief inducement for the many strangers who visit it from Innsbruck is the very fine view of the valley obtained from the Schlossthurm. The rly. is carried along the l. bank till it passes the junction of the Sill with the Inn opposite Mublau ; it then crosses the river and enters by a viaduct the city of

Innsbruck (described in § 42, Rte. A).

ROUTE C.

MUNICH TO BRIXLEGG IN THE INNTHAL, BY SCHLIERSEE.

Railway to Miesbach, about 29 English miles — Miesbach to Neuhaus, carriage road ; thence to Brixlegg, foot path; about 12 stunden or 36 English miles—in all about 65 m.

Though not presenting so much variety as the road by Kreuth and the Achensee, described in Rte. A, the unfrequented way to the Innthal here noticed has many attractions for the lover of nature.

At Holzkirchen (Rte. A), a short branch rly. leads to *Miesbach* (Inns: Post; and several others), a pretty country town frequented by visitors from Munich. The castle of Wallenburg and the summits of the neighbouring hills offer pleasing views. An omnibus conveys passengers in one hr. from the rly. station to the *Schlier See*, a charming little lake, one of the gems of the Bavarian Alps, little known to foreigners, but long a favourite resort of Bavarian summer visitors. Etymologists derive the name from the fish Silurus, for which the lake was famous among Roman epicures. Visitors find good accommodation at the village of *Schliersee* (Inns: Bei der Fischerliesel; Post). Comfortable lodgings are found in the clean and neat farm-houses of the village and neighbourhood. Though on a very small scale, for it is but 2 m. long, and the mountains little more than 3,000 ft. above its shores, this little sheet of water has beauties of its own that are not easily matched elsewhere. It may be easily visited from Tegernsee (Rte. A), with which it is connected by a carriage-road (passing Agatharied and Gmünd), and by several mountain paths. The shortest of these—to be preferred in going from Tegernsee to Schliersee—is by the *Kreuzalp*, a low and easy pass, quite practicable for ladies—distance only 3 hrs. The view in descending towards the Schliersee is very beautiful. A longer way, commanding much more extensive views, is by the *Gindelalp* (4,721'), lying farther N., and overlooking a wide extent of the Bavarian plain. The favourite excursion from Schliersee is to the ruined castle of *Hohenwaldeck*, standing on a projecting rock (3,273'), immediately above the village. According to local antiquaries the foundations date from the pre-Roman period. It was the seat of the ancient family of Waldeck, which became extinct, in the male line, in 1483.

The carriage-road along the E. shore traverses a low pass S. of the lake, and then turns eastward to follow the Aurach. one of the branches of the Lietzach (Rte. D). At a solitary inn called Neuhaus the traveller bound for the Innthal leaves the road to follow a branch of the stream that descends from S. to N. On passing through a short defile he unexpectedly comes on the little village of *Max-Josephsthal*, founded in the last century by the last Count of Hohenwaldeck, whose name it bears. Mounting through the glen (called Josephsthal), a path leads to the Stockeralp (about 4,000') forming the watershed between the stream running S. to the Inn and the above-mentioned tributary of the Leitzach. The pastures on the S. side of the pass are called Spitzingalp, and the traveller soon reaches the *Spitzingsee* (3,542'), a pretty little lake surrounded by green slopes. This is the source of the Rothe Falep, or Spitzingbach, and lies at the N. extremity of the *Brandenburgerthal*, one of the few considerable valleys that run from N. to S. towards the valley of the Inn. About 1¼ hr. below the lake the Rothe Falep is joined by the Weisse Falep, flowing from WNW. A path descends along that stream from the height of land at the N. side of the Schinder (Rte. A). Thence Tegernsee may be reached by following the stream of the Rothach, and another path descends through the Langenauthal to Wildbad Kreuth. Less than 1 m. below the junction of the two Faleps the glen is narrowed to a defile, and a few hundred yards N. of the Tyrolese frontier the traveller sees the remains of the Kaiser Klause. This was a massive dam by which the course of the stream was barred at pleasure until a considerable mass of water was accumulated behind it, which was used to transport the vast supplies of timber cut down at the head of the valley to feed the smelting furnaces at Brixlegg and Rattenberg. On removing the barrier the pent-up waters rushed down the valley, carrying with them large quantities of floating timber, but leaving a large portion on the way arrested by obstacles along the banks. A similar method is commonly adopted elsewhere in the Alps, but not often on so large a scale. The Kaiser Klause has been abandoned since 1827, when a new dam was constructed lower down, some way within the frontier of Tyrol. It is called Erzherzog-Johanns-Klause, and is of still larger dimensions. The opening of the barrier generally takes place during the early part of the summer, and is a sight worth seeing. There is a mountain-path leading hence to Kreuth by the E. end of the Planberg. The portion of the valley lying between the old and new Klause is a cleft between that ridge and the *Hinter-Sonnenwend-joch* (6,472′), which seems to form an eastern prolongation of the same range. The way now follows the somewhat sinuous course of the Brandenbergerthal chiefly along the l. bank of the torrent—here called *Bayerbach*, as flowing from Bavaria. About 2 hrs. below the new Klause a glen opens eastward, through which runs a path to Thiersee (Rte D), and nearly opposite to it is the *Steinbergerthal*, leading westward in 7 hrs. to Achenkirch (Rte. A), by the village of Steinberg, and a pass on the N. side of the Unnütz. The chief hamlet of *Brandenberg* stands on a mountain terrace, 2,950 ft. above the sea. If bound for Wörgl the traveller should follow a path to the eastward, which will take him to the l. bank of the Inn about 4 m. above that village. More interesting is the path along the main stream, here called Brandenberger Ache, through a fine defile which opens out at *Mariathal*, one of the five hamlets making up the commune (gemeinde) of *Voldepp*, which, along with many scattered farm-houses, occupy a sort of recess on the N. side of the Innthal. ENE., parallel to the course of the river, extends the *Mooserthal*, a marshy tract with six shallow lakes, occupying a portion of the broad trough through which the Inn escapes into the plain of Germany. The neighbourhood of Voldepp is rich in marbles, used in the churches of Innsbruck and the neighbouring villages. Voldepp is about 1 m. from Rattenberg, and the same distance from the rly. station at Brixlegg (Rte. B).

Route D.

MUNICH TO KUFSTEIN, BY BAIERISCH-ZELL.

Railway to Miesbach, about 29 Eng. miles; carriage road to Baierisch-Zell, about 19 Eng. miles; footpath to Kufstein about 7 stunden, or 21 Eng. miles. Instead of following the railway to Kufstein (Rte. B) some travellers may prefer to see something of an attractive but very unfrequented portion of the Bavarian highlands.

The chief valley that penetrates the limestone ranges on the frontier of Bavaria and Tyrol, between the Inn and the Mangfall is that of the Leitzach, and it is that naturally chosen for a mountain route to Kufstein. The Leitzach is traversed by the post road from Miesbach to Rosenheim, and a hilly branch road then turns S., passing *Ellbach*, *Marbach* (a good country inn, once a nobleman's house), and *Fischbachau*, always keeping at some distance eastward of the stream, till it crosses to the l. bank to gain *Aurach*, near to which village the Leitzach receives, from the W., the torrent also named Aurach. The shortest as well as the more interesting way from Miesbach to Aurach is, however, that by the Schlier See, mentioned in the last Rte. Above Aurach the valley of the Leitzach mounts SE. between the *Wendelstein* (5,992′) and the *Miesing* (6,042′); it contains many scattered houses but no villages. The principal place, named in memory of the Benedictine convent which once existed there,

Baierisch-Zell (2,627′), is a group of only a dozen houses with an inn. Both the above-named mountains may be ascended from hence, but, although a few ft. lower, the Wendelstein, lying farther N. between the Leitzach and the Inn, commands the finer view. It is said that the ascent from Fischbachau is more interesting. The summit might be taken on the way from Marbach to Baierisch-Zell, but not without a guide. The pedestrian who would follow a less laborious course without making the detour to Aurach may take a path on the rt. bank of the Leitzach leading to Zell, in 3 hrs. from Marbach. The valley of the Leitzach seems to be closed by the mountains that surrounded Baierisch-Zell, but the chief source of the stream descends due N. through a narrow glen. Except after heavy rain the bed of the torrent is dry, and the stream finds its way underground from the *Stocker Seen*, two inconsiderable lakes which are considered to be its proper sources. The Pass of *Hörhag* lies just beyond, and very little above, the lakes, and, as often happen on its N. frontier, the boundary of Tyrol lies somewhat S. of the watershed. The first Tyrolese village is *Landl*, reached in 3 hours from Baierisch-Zell. This lies at the upper end of the *Thierseerthal*, a glen drained by the Thierseer Ache, which runs somewhat N. of E. till it joins the Inn on the frontier of Tyrol and Bavaria, close to the village of Kiefersfelden (Rte. B). The names of the glen and of the two villages, Inner- and Ausser-Thiersee are taken from a little lake, *Thier See*, finely situated in the lower part of the glen. Instead of following the course of the stream, the shortest and most interesting way to Kufstein is over the ridge dividing the Thierseerthal from the Inn. The path descends the steep E. slope of the *Pendling* (4,992′), partly through forest, with occasional fine views over the Innthal.

ROUTE E.

TEGERNSEE TO MITTENWALD, OR PARTENKIRCH.

The traveller who would become acquainted with the finest scenery of the Bavarian Alps will endeavour to combine a visit to Tegernsee and Kreuth (described in Rte. A) with the district surrounding Partenkirch and Mittenwald, which was noticed in the last section of this work. There are various roads and mountain paths by which this object may be effected, and the more useful of them are here briefly noticed.

1. *By post road through Tölz.* 5 Bavarian, or 23 English miles to Benedictbaiern, the same distance from that place to Mittenwald. Those who wish to keep altogether to the post road must follow the road from Tegernsee towards Holzkirchen (Rte. A.), till it intersects that which runs nearly due W. from Rosenheim to Benedictbaiern. Turning into this line of road, the traveller passes several villages and a massive building, formerly the monastery of Reichersbaiern, now used as an agricultural institution. He reaches the banks of the Isar at

Tölz (Inns: Bürgerbräu; Post), a small town, finely situated on rising ground, 2,551 ft. above the sea, on the rt. bank of the river. The views of the Isarthal from the garden of the best inn, and from the Calvarienberg are very pleasing. The mineral springs of *Krankenheil* and *Bocksleiten*, containing iodine, sulphur, and alkaline salts are on the l. bank of the Isar about 1 m.

from the town. The Zollhaus has been opened there as an inn and bathing establishment. The range of the Benedictenwand is frequently in view to S., during the stage of 2 Bavarian m. between Tölz and Benedictbaiern, where this road joins that from Munich to Walchensee and Mittenwald, described in § 42, Rte. G. An omnibus plies daily between Tölz and Benedictbaiern, corresponding with that which connects the latter place and Seeshaupt on the l. of Starnberg (§ 42, Rte. D).

2. By the *Jachenauthal*. Instead of following the post road from Tölz to Benedictbaiern and thence to Walchensee, there is a way 6 or 7 m. longer, but much more interesting, by a country road through the valley of the Isar, and along the stream that flows into that river from the Walchen See. An *einspänniger wagen* from Tölz to Walchensee (about 27 m.) costs about 5 Bav. florins. The way lies along the level rt. bank of the Isar for about 7 m. from Tölz to

Länggries (with two Inns), the only considerable village in the valley between Tölz and Mittenwald. S. of the village the Isar flows through one of those depressions or clefts in a transverse ridge of mountain so characteristic of the orography of the Suabian Alps. The road crosses the Isar by a massive wooden bridge opposite the village, and keeps for about 3 m. more to the l. bank of the Isar. The modern castle of *Hohenburg*, standing near the ruins of an ancient castle so-named, is the most conspicuous object in the valley. It belongs to Baron Eichthal, the banker, of Munich.

At the point where the road turns aside from the Isar towards the Walchen See, stands a decent country inn, called Pfaffenstöffel. Here the *Jachenauthal* opens to WSW., and the road is carried along the l. bank of the Jachen which drains the Walchensee into the Isar, receiving several minor streams from the mountains on either side. The sparse population, said to offer the finest specimens of the vigorous highland race of Bavaria, is scattered in single houses or small groups in the comparatively few open spaces not occupied by the aboriginal forest. The 'Schnaderhüpfeln,' a kind of improvised dialogue sung between two contending performers, is said to be heard here in perfection at festive meetings. The church and principal hamlet, with an inn, are at *Jachenau* (2,427'). Here the road leaves the Jachen, and mounts nearly due W. across a ridge, descending thence to Sachenbach on the N. shore of the Walchensee. About 2 m. farther is *Urfeld*, where the cross road hitherto followed joins the high road from Benedictbaiern to Walchensee and Mittenwald (§ 42, Rte. G).

3. *By the valley of the Isar.* There is probably not one of the considerable streams flowing from the Alps that sees so slight traces of the presence of man upon its banks as the Isar throughout the long space (12 stunden, following the track) between the above-mentioned village of Länggries and Mittenwald. For fully 25 miles there is not a hamlet and scarcely a house to be seen, except a few saw-mills, and the dwellings of gamekeepers. The absence of population has of course favoured the increase of wild animals, and game of every sort, including chamois, is said to be abundant. The scenery of the lateral glens, mentioned below, is more interesting than that of the main valley, but to some travellers the quiet and seclusion of this route may offer no slight attraction. The traveller starting from Tölz may take a light one-horse vehicle as far as the 'Fall,' 6 stunden, or 18 m.—fare 3½ Bav. florins.

[The pedestrian wishing to join this route from Wildbad Kreuth may follow the road up the glen of the Weissach (Rte. A) as far as Stuben. A track over the ridge westward leads thence to the valley of the Isar a little above the junction of the Jachen with that river.]

Instead of crossing the bridge at Längries as on the way to Jachenau, the char road from Tölz keeps to the rt. bank of the Isar. Before long the valley begins to bend to W., and gradually takes a WSW. direction parallel to that of the Jachenauthal, from which it is separated by a ridge of moderate height, traversed by several forest paths. The first break in the somewhat uniform scenery of the valley is at the affluence of the stream from the Achensee, called simply Ache in Tyrol, but *Walchen* on this side of the frontier. Beyond this point the valley is contracted between the base of the Hennenkopf to N., and the Dürrenberg on the S. side of the stream, which at the narrowest point forms rapids 15 ft. in height, presenting a serious difficulty to timber-rafts. The road soon after reaches Fall, a small group of houses above the rapids, at the junction of the *Dürrach*. This torrent, formed by the confluence of the waters falling into a wild mountain amphitheatre on the Tyrolese side of the frontier, is further noticed in the next route. The mountaineer may make an interesting detour by ascending nearly to the head of the glen, crossing the ridge on the N. side of the *Scharfreiterspitz* (6.848'), and returning to the valley of the Isar through the Rissthal.

The char road formerly came to an end at *Fall*—6 stunden (18 Eng. m.) from Tölz, where comfortable quarters and excellent food are found at an inn frequented by sportsmen. A good new road has been extended 2½ stunden farther up the valley to *Vorder-Riss*, at the junction of the Rissthal (Rte. F) with the Isar. Here stands a shooting lodge, belonging to the late King Max, and a keeper's house—Forsthaus—now opened as an inn, very fair quarters, but not cheap. The portion of the Isarthal above Vorder-Riss is the most lonely of the entire valley. For about 8 m. the pedestrian must follow the track along the river before he sees any trace of human activity. At last the valley bends to S., and he joins the road from Walchensee, where it approaches the l. bank of the Isar at the village of *Wallgau* (§ 42, Rte. G). A rather shorter but rougher way is by the path along the rt. bank of the Isar. If bound for Partenkirch he should choose the former course, turning off from the high road at Krün. The distance from Vorder-Riss to Mittenwald is counted 5¾ stunden or 17¼ Eng. m.

Instead of following the Isar by the direct track, the pedestrian may take a much more interesting, and not much longer route, through the *Fischbachthal*, a glen which, running parallel to the Isar, joins the Rissthal about 2 m. above Vorder-Riss. The upper end of the glen is encircled by bold limestone peaks, the highest of which is the *Soiernspitz* (7,303'). In the midst of these are three small lakes, *Soiernseen*, with a Sennhütte beside the highest of them. Schaubach recommends the traveller bound for Mittenwald to return for ½ hr. by the same path which led him to the lakes, and then turn northward over a low ridge called Fishbachalp, whence a path leads him down to the Isar opposite Wallgau; but it is doubtless a more interesting though a more laborious course to cross the ridge SW. of the lakes, and descend thence to Mittenwald. Another, and still longer way from Vorder-Riss to Mittenwald is to ascend the Rissthal as far as the junction of the Fermersbach, the second stream that joins the Riss from WSW., having, like the Fischbach, a course parallel to that of the Isar. Mounting through the *Fermersthal*, which here marks the frontier between Tyrol and Bavaria, a path crosses the ridge of the Vereinalp at its head, descending immediately under the bold precipices of the *Karwändlspitz* (8,259') which are here seen to the utmost advantage. The path to Mittenwald joins that above noticed from the Soiernseen.

Route F.

WALCHENSEE TO PERTISAU ON THE ACHENSEE.

The roads and paths mentioned in the last Rte. all follow a general direction from NE. to SW. parallel to the frontier of Tyrol and Bavaria, but keeping within Bavarian territory. In the present Rte. we shall briefly notice the tracks traversing the same district in a transverse direction from NW. to SE., assuming that some lovers of alpine lake scenery may wish to pass from the milder slopes that encompass the Bavarian Walchen See to the sterner shores of the Tyrolese Achensee.

1. *By the valley of the Ache.* Having reached the hamlet of Jachenau either by the char road (Rte. E), or by taking boat from Walchensee to Sachenbach, the traveller may either follow the Jachenauthal to its junction with the Isar, and then remount along the latter stream to its junction with the Walchen (or Ache), or else follow, with a guide, a path leading from Jachenau over the ridge a little S. of E. to Fall on the Isar. If the writer is correctly informed that there is no bridge at the opening of the Jachenauthal by which a light carriage can cross the Isar and join the road from Länggries to Fall, it is certain that the second of the above alternatives is the shorter course. Having entered the valley of the Isar at or near to Fall, the shortest way to the Achensee is by the valley of the Ache, which drains that lake into the Isar. The road from Kreuth, described in Rte. A, enters that valley a short way above the point where the Ache enters Ba-varia and assumes the name *Walchen*. In its short course on that side of the frontier the Walchen has cut a deep ravine through which it descends to the lower level of the Isar. A pretty good new road replaces the horse path constructed for the late King Max, who often visited this district for chamois-hunting, and by that way the village of Achenkirch is reached in 3 hrs. from the junction of the Walchen with the Isar. It might be possible to reach Seehaus, or even Pertisau (Rte. A), in one long day from Walchensee, but, considering the uncertainty of finding room at either place, the traveller will do well to halt for the night at Achenkirch, or at Kern's inn ¼ hr. before reaching that village.

2. *By the Dürrach, or Pfanserthal.* Having reached Fall by either of the courses mentioned above, the most direct and probably the most interesting way to Pertisau is by the glen of the *Dürrach*, which there joins the Isar. This is sometimes called *Pfanserthal*. About 2 hrs. above Fall the traveller enters Tyrol, and keeping on due SW. he will in 1½ hr. more find himself in the centre of a complete circle of bold limestone mountains. Maintaining the same direction a faintly marked track will lead him to the pass of the *Pfans Joch*, between the Rabenspitz and the Keelberg. He descends thence into the *Gernthal*, the northernmost of the two mountain glens that unite above Pertisau. At least 7 hrs. should be allowed for the walk from Fall to that place.

3. *By the Rissthal.*—The Rissthal is a long, almost uninhabited glen a paradise of sportsmen—wh.ch, like the main valley, originates on the Tyrolese side of the frontier, and enters Bavaria 5 or 6 miles above its confluence with the Isar. As mentioned in Rte. E, the traveller will find good quarters at the Forsthaus at *Vorder-Riss*. He may cross the lake from Walchensee to Altach, whence a well-marked track crosses the Hoch Schott, and enters the Isar valley about ½ hr. above Vor-

der-Riss, or go by Jachenau, and by a short pass leading southward from that place to Vorder-Riss. For a distance of 3 hrs the Rissthal extends due S. nearly at a level to *Hinter-Riss*, within the Tyrolese frontier, where the Duke of Coburg Gotha has a rather large shooting-lodge, built in modern Gothic style. A little inn has been opened by a gamekeeper, and supplies fair accommodation. The scenery of the Upper Rissthal is extremely fine, and that of its lateral glens, especially the Blaubachthal, rises to grandeur. Chamois and red deer, rarely disturbed, are often to be seen at a low level. Among other rare plants, *Astrantia gracilis* is common on the limestone mountains between this and the Achensee. The naturalist who may halt a few days in this retired glen will find here ample occupation, and several interesting mountain excursions may be made from this point. To NE. is the *Scharfreiterspitz* (6,848'), commanding a fine view. On the W. side of the Rissthal are the two lateral glens *Fischbachthal* and *Fermersthal*, each of which will afford scope for an interesting excursion. A traveller entering the Rissthal from Mittenwald would naturally choose one or other of the passes mentioned in Rte. E, in preference to the dé·our by Vorder-Riss.

From Hinter-Riss, where the valley bends to ESE., the path ascends for 2 hrs. to the Hagelhütte (3,350'). Then bearing to l., a rapid ascent for 2 hrs. more leads the traveller to the summit of the *Plumser Joch* (5,492'), where a large chálet supplies milk and butter. In 1½ hr. he may easily descend from the summit through the Gernthal to Pertisau. A rather longer way through very grand scenery is to turn to the rt. at the head of the Rissthal into the *Blaubachthal*, which stretches for some miles nearly level between precipitous crags, and then ascend to the l. by the first sennhütte. A newly made path makes the way easy. By keeping to the l., where there is a choice of ways, the traveller will reach the *Garmey*- joch overlooking the head of the Falzthurnthal, and descend that way to Pertisau.

ROUTE G.

SCHARNITZ TO JENBACH, IN THE INNTHAL.

In describing the Alpine region that lies between the Lake of Constance and the lower course of the Inn, the writer has had to point out the frequent recurrence of parallel ridges of sedimentary rocks preserving a direction that varies little from due E. and W.; and it was seen that the line from Bludenz in the Vorarlberg to Hall below Innsbruck, which maintains the same direction, is fixed by nature—on orographic as well as geological grounds—as the southern boundary of this region. An important road, described in preceding portions of this work, extends along the valleys connecting those two places, but it has also been seen that the existence of numerous minor parallel valleys makes it easy to devise routes in the same direction by which the main road is altogether avoided, without encountering the numerous obstacles that might be anticipated. Thus a traveller starting from Feldkirch, following the Walserthal, and the Lechthal as far as Reutte, thence taking the road to Lermos, and proceeding eastward by the Gaisthal, may reach Scharnitz (§ 42, Rte. D) without encountering any high transverse ridges, and without diverging widely from a straight line. At Scharnitz the traveller has a choice of routes by which to continue his journey in the same direction, but inasmuch as the Innthal bends towards NW. between Hall and Kufstein, whichever of the easterly lines he may pursue he will ultimately intersect the main valley at some point between Hall and Rattenberg.

The three paths which offer themselves to the mountaineer at Scharnitz follow the three Alpine torrents whose union forms the Isar near that village. Each of them passes amidst very fine scenery, almost completely unknown to travellers, who are merely aware in a general way that there is a group of high mountains N. of the Inn near Innsbruck.

1. *By the Karwändlthal.* The northernmost of the torrents forming the head waters of the Isar flows through the *Karwändlthal*, lying on the S. side of the range of the *Karwändlgebirge*, and separated from the Hinterauthal noticed below by the Riedlkar range, most of whose summits considerably exceed the limit of 8,000 ft. Although the Karwändl, like the ranges noticed below, is essentially an E. and W. ridge, it throws out a considerable spur towards SW. in the direction of Scharnitz, along which runs the frontier line of Tyrol and Bavaria. This spur turns the torrent to the S., and forces it to join that from the Hinterau about ½ hr. above Scharnitz. The path mounts by the rt. bank of the torrent along the base of the Karwändl range, whose flanks are defaced by vast slopes of débris. After passing below the highest summit of the range—the *Karwändlspitz* (8,259')—the path mounts rapidly to the Hochlaner Alp. Here the Karwändl range comes abruptly to an end, while the parallel range to S. extends eastward for many miles to the *Grubenkorspitz* (8,755'), and the *Vomperjoch* (7,505'). A small chapel marks the summit of the pass, and the ground falls rapidly northward towards the Rissthal. Here the traveller may descend a little towards that valley, then turn to rt. and cross the Garmeyjoch (Rte. F), to Pertisau; but a good local guide can lead him by a somewhat intricate course along the flanks of the *Birkkorspitz* (8,978'), and by the northern face of the Grubenkor, till he reaches a ridge called Kaisergrat, whence he may either descend to Pertisau by the Falzthurnthal, thence reaching Jenbach by the high-road, or follow a more direct course to the valley of the Inn through the Stallenthal, passing the Georgenberg (Rte. B), and reaching the road at Stans, about three m. above Jenbach.

2. *By the Hinterauthal.* The torrent from the *Hinterauthal*, opening due E. of Scharnitz, is generally accounted the main source of the Isar. The glen is walled in on the N. side by a ridge, which may be called the Riedlkar range from its most conspicuous summit as seen from Scharnitz, though the eastern end of the same range, extending eastward beyond the basin of the Isar, attains a greater height in the Birkkorspitz, mentioned above. In the same manner the range of the Gleirsch Joch, dividing the glens of Hinterau and Gleirsch, attains its highest point at its E. extremity in the *Lavatscherspitz* (9,081').

The Hinterauthal is a wild glen, resorted to only by a few herdsmen in summer. It serves however as the most direct way from the neighbourhood of Scharnitz to the lower Innthal. A tolerable path mounts along the rt. bank of the torrent, and after passing near to a spring, which is called the source of the Isar, reaches the pass of *Haller Anger* (5,835'), separating this from the *Vomperthal*, which falls in the opposite direction due E. towards Schwaz. This is said to be one of the wildest and most striking glens of this district, especially towards the lower end, where it is narrowed to a savage defile which opens out above the village of *Vomp*, near to the ruined castle of Sigmundslust. The village, with a pretty church containing a picture by Albert Dürer, was destroyed by the French under Deroi in 1809. About 1 m. farther is Schwaz (Rte. B).

3. *By th Gleirscherthal.* This is the southernmost of the three glens intervening between the Bavarian frontier and the valley of the Inn. As the torrent from the Karwändlthal was bent southward by a spur from the ridge to the N., so that issuing from

the Gleirscherthal is turned northward by a similar spur extending NW. from the range of the Solstein. After ascending for some distance to SE. the head of the glen turns due E. Here the mountaineer may well be tempted in fine weather to ascend the *Solstein*, sleeping the night before in one of the sennhütten near its base. The effect of the view from the top must be much enhanced when it is suddenly opened before a traveller arriving from the N. side, and the descent towards the Inn valley (§ 42, Rte. A) will be less laborious than the ascent from Zirl. The course formerly taken from the Gleirscherthal to Innsbruck by crossing the ridge E. of the Solstein is said to have become very difficult, or impossible, owing to a fall of rocks.

The track to the Lower Innthal mounts about due E., passes a solitary shooting-lodge, 3 hrs. from Scharnitz, and in as much more time reaches the *Stempeljoch* (7,346'), the pass dividing this glen from the Hallthal. Hence the traveller may follow the stream down to the main valley at Hall, or else, if bound for Schwaz and Jenbach, turn aside at Baumkirchen, and enjoy an agreeable woodland walk through the Gnadenwald (Rte. B). Further information as to the passes described in this and the last Rte. is much desired.

CHAPTER XIII.

SALZBURG ALPS.

Section 44.
KITZBÜHEL DISTRICT.

Route A — Innsbruck to Salzburg, by Lofer 56
Route B — Munich to Salzburg, by railway 62
Route C — Wörgl to Mittersill, in Pinzgau, by Kitzbühel . . 63
Route D — Hopfgarten to Wald, in Pinzgau 65
Route E — Traunstein to Kitzbühel . 66
Route F — St. Johann to Saalfelden, by Fieberbrunn . . . 69
Route G — Kitzbühel to Zell-am-See, by the Glemmthal . . . 70

Section 45.
BERCHTESGADEN DISTRICT.

Route A — Munich to Berchtesgaden, by Reichenhall . . . 72
Route B — Salzburg to Zell-am-See, by Berchtesgaden . . . 82
Route C — Salzburg to Zell-am-See, by Reichenhall . . . 86
Route D — Berchtesgaden to Saalfelden, by the Steinerne Meer . 87
Route E — Salzburg to Lend, by the valley of the Salza . . 90
Route F — Berchtesgaden to the Valley of the Salza . . . 95
Route G — Saalfelden to Lend, by the Urschlauthal . . . 96

Section 46.
ISCHL DISTRICT.

Route A — Salzburg to Ischl—Excursions from Ischl 98
Route B — Linz to Ischl 109
Route C — Salzburg to Gmunden or Lambach, by the Mond See and Attersee 112
Route D — Ischl to Steinach, in the Ennsthal, by Aussee . . . 114
Route E — Ischl to Golling, by Gosau—Ascent of the Dachstein . 116
Route F — Abtenau to Radstadt or Werfen, by the Fritzthal . . 122
Route G — Abtenau to Werfen, by the Tännen Gebirge . . . 123

Section 47.
ENNS DISTRICT.

Route A — Enns, on the Danube, to Radstadt, by the Ennsthal . 125
Route B — Steyer to Lietzen, by the Pyrhn Pass . . . 132
Route C — Dürrenbach to Aussee, by the Todtes Gebirg . . 134
Route D — Dürrenbach to Ischl or Gmunden 137
Route E — Windischgarsten to Altenmarkt, in the Lower Ennsthal 138

THE writers who have described systematically the Eastern Alps have differed but little in respect to the boundary between the northern and the central divisions of the chain. It has been seen in the last chapter, that as regards the western half of the region in question, the valley of the Inn from Hall to Landeck, and the valleys traversed by the road from Landeck to Feldkirch, form a boundary, which is equally fixed by the orography and the geology of the country. Below Hall, however, the Inn valley bends to NW., and after passing Wörgl, it no longer marks the division between the crystalline and sedimentary rocks, as it fairly enters the zone formed by the latter. At the same time a glance at the map suffices to show another great line of valley, extending in a line nearly due E. from Hall, which approaches to, though it does not actually reach, the valley of the Inn. A straight line, 120 m. in length, drawn from the elbow of the Enns below Admont (where that stream turns N. to join the Danube) to the head of the Pinzgau, is nowhere more than 2 or 3 m. distant from one or other of the streams which flow through that long trough; and the same line of depression is continued westward over the low Gerlos Pass to Zell, in the Zillerthal. For more than half its length this trough marks the limit between the secondary limestones and the crystalline rocks of the central chain. Along its western portion the geological significance of the same boundary has been less evident, for between Lend and the Zillerthal a broad zone of crystalline slates is seen on the N. side of the Pinzgau. But when it is considered that the N. portion of this zone exhibits triassic strata overlying slates that have been called Devonian, but are now referred to the silurian age, it will not appear unlikely that the remaining portion of the zone may be ultimately referred to the latter epoch; the rocks differing only as to the degree of metamorphic action which they have undergone.

It must be observed that there is a second line of depression, nearly parallel with the first, but less well defined orographically, and less straight, which marks very nearly the division between the jurassic limestone mountains to the N.. and the arenaceous and schistose strata of the intermediate zone. This line is marked by the Kaiserstrasse road from Wörgl to St. Johann (§ 44, Rte. A , by the track leading thence by Hochfilzen to Saalfelden (Rte. F), and by the Urschlauthal and Mühlbachthal connecting the latter town with St. Johann im Pongau. Hence some writers have thought it preferable to include the intermediate region, whose centre is Kitzbühel, among the minor groups of the central chain.

In the present work orographic conditions must prevail over every other, and the writer does not hesitate to follow the example recently set by V. Sonklar, and to regard the main valley of the Pinzgau as the natural division between the northern and central divisions of this portion of the Eastern Alps.

But while maintaining that the region lying N. of the limit above defined, and between the Inn and the Enns, as its western and eastern boundaries, forms a natural and well-defined division of the Alps, it is not easy to find for it an appropriate designation. The circumstance that it includes portions of Bavaria, Tyrol, and of the three Austrian provinces of Salzburg, Styria, and Upper Austria, creates an objection to the name which it seems reasonably to take from the ancient city of Salzburg, which is the central point and natural capital of the region in question. The designation 'Salzburg Alps' has indeed been often applied to some portion of the region now in question, and C. von Sonklar, the latest and most accurate writer who has remodelled the divisions of the Eastern Alps. has applied it to the small district between the Salza and the Saale, the greater portion of which belongs to Bavaria. Feeling the difficulty of suggesting any other name generally acceptable, the writer, for the purposes of this work, extends the name Salzburg Alps to the entire region included within the above-defined boundaries.

Some portions of the region here to be described are already well known to travellers. The neighbourhood of Berchtesgaden, and the lake district of the Salzkammergut, are not surpassed, and can scarcely be equalled elsewhere, for a character of scenery in which variety is the main characteristic. Intermediate in many respects between the lakes of our own islands and those of Switzerland and Italy, these far surpass the first in dimensions and in the bold outlines of the surrounding mountains, while they make no pretensions to the stately grandeur of the last. From both they differ in the more frequent and more vivid contrasts of colour and form. Nowhere else in the Alps does the traveller so frequently pass, in the course of a few hours' walk, from scenes of soft beauty associated with

§ 44. KITZBÜHEL DISTRICT. 55

the signs of human industry and comfort—not seldom with the traces of elegance and luxury—to the savage sublimity of rugged glacier-clad peaks.

To the mountaineer the group of the Dachstein, although it nowhere quite attains the height of 10,000 ft., offers, besides other attractions, the excitement of difficulty. Several of its peaks are yet untouched, and all must be reckoned as decidedly difficult of access.

Besides Ischl, which supplies first-class accommodation to its courtly visitors, there are many spots noticed in the following pages where comfortable quarters are found; and even in the more unfrequented places a stranger is generally secure of a clean bed and tolerable food.

been already indicated. To the W. the Inn and the Ziller as far as Zell; to S. the Gerlos Pass, and the upper valley of the Salza; to E. the Saale from its junction with the Salza to Saalfelden, and the road leading thence to the Pinzgau—these boundaries circumscribe the present district save to N., where it subsides into the Bavarian plain. Though far from having attained the celebrity which the two succeeding districts have acquired through the beauty of their lake scenery, this is not deficient in natural attractions; and, either because of its natural richness, or because it has been more carefully examined, the neighbourhood of Kitzbühel attracts botanists by an unusual number of rare plants. Very good quarters are found both at Kitzbühel and Lofer, and there are tolerable country inns in most of the villages.

The lower Zillerthal and the upper Pinzgau, although lying on the boundary of this district, are more conveniently described in connection with the central chain of the Tyrol Alps; but on the other hand, it seems advisable to include the city of Salzburg, although it lies a few miles eastward of the Saale; inasmuch as this district is mainly known to strangers as being traversed by the important road from Innsbruck to Salzburg.

According to the recent government survey the highest summits of this district are the Thorhelm (8,548'), the Katzenkopf (8,311'), at the head of the Kelchsauerthal (Rte. D), the Birnhorn (8,635'), and the Mitterhorn (8,326'), with the adjoining peaks of the Lofer Alps.

SECTION 44.

KITZBÜHEL DISTRICT.

The designation 'Kitzbühel Alps' has been applied by several German writers to the range, alluded to in the introduction, which lies between the head of the Pinzgau and the road leading from Wörgl to Saalfelden. In this work the mountain tract extending on the N. side of the same range, between the Inn and the Saale, has been included along with it in this district, named after the little town of Kitzbühel, perhaps more correctly written Kitzbüchl, which lies near its centre. The boundaries have

ROUTE A.

INNSBRUCK TO SALZBURG, BY LOFER.

	Austrian miles.	English miles.
Wörgl (railway) . .	8	37¾
Söll (road) . . .	1¾	8¼
Elmau	1½	7
St. Johann . . .	2	9½
Waidring . . .	2	9½
Unken . . .	2½	11¾
Reichenhall . . .	2½	11¾
Salzburg (railway) .	2	9½
	22¼	105

Travellers pressed for time will naturally avail themselves of the railway between Innsbruck and Salzburg, but in so doing they lose much fine scenery which they would enjoy by the present Rte.; besides which, this is naturally combined with a visit to Berchtesgaden and the beautiful Königs See, described in the next §. The most convenient stopping-place on this road is at Waidring; but those who wish to halt and enjoy the finest scenery of the district at their leisure will prefer Lofer.

Allowing for delays, at least 12 hrs. are required for posting from Wörgl to Reichenhall.

According to the latest information a *stellwagen* starts daily at 11 A.M. from Wörgl, in connexion with the morning train from Innsbruck, and reaches Lofer at about 6 P.M. Going on again next morning, it reaches Reichenhall at 9 A.M. In the opposite direction it leaves Reichenhall at midday and stops for the night at Waidring.

After crossing the bridge over the Brixenthaler Ache, the road quits the valley of the Inn below a projecting eminence called Grattenberg, and is carried for 4 m. along the rt. bank of the former stream to the western foot of the Hohe Salve. The principal branch of the Brixenthal opens to SE., but the road to Salzburg ascends NE. through a short and narrow lateral glen. Soon attaining the watershed, the road descends very slightly before reaching

Söll (Inn: Post), the chief village of the short but very picturesque vale of *Sölland*, drained by one of the numerous streams called Weissach, which joins the Innthal a little above Kufstein (See § 43, Rte. B). To NE. is seen the limestone group of the *Kaisergebirge*, whence is derived the name Kaiserstrasse, often given to this road. The bold precipitous ridge projecting westward from the main mass is called Mosberg, and near the foot of the rocks is the *Hintersteiner See*, overlooked at its E. end by the church of Bärenstatt: the circuit of the lake affords a pleasant stroll from Söll. The ascent of the Hohe Salve (Rte. C) may be made in 3 hrs. from the village, but it is advisable to take a guide as far as the first châlet; thence it is not easy to miss the way. By bearing to rt., near the top, the traveller will fall into the bridle-path from Hopfgarten; but it is a better, though a steeper, course to steer straight for the summit. *Scheffau* (2,384′) stands near the S. base of the principal group of the Kaisergebirge, whose highest peak—the *Scheffauer Kaiser* (7,611′)—is reached in about 5 hrs. from the village. The last rocks are very steep. About 3 m. farther the road attains the water-shed between the Weissach and the Kitzbühler Ache, 2,840 ft. above the sea, and very soon after reaches the post-station at

Elmau (Inn: Post), a small village forming the eastern limit of the district of Sölland. The road now descends along the l. bank of the Rheinach to

St. Johann (Inns: Post, good, civil landlord; Schwarze Bär, a well-looking house opposite the post), a village standing 1,950 ft. above the sea, at the junction of several streams, each flowing through a glen traversed by a road. Besides the Rheinach, by which our road descended from Elmau, the main stream of the Kitzbühler Ache flows from S., and by that way a carriage road reaches Kitzbühel, only 7 m. distant, there joining the route to Mittersill (Rte. C). Another glen, opening ESE., is traversed by the road to Saalfelden (Rte. F). For a further notice of the Kaisergebirge see Rte. E.

Near St. Johann is the abandoned

mine of *Röhrerbühel*, which, after producing large quantities of silver and copper, was abandoned towards the end of the last century. It was considered at one period to be the deepest mine in the world, the shaft having been carried to a depth of 3,128 ft. below the surface.

The road to Salzburg follows for 5 m. in a NNE. direction the rt. bank of the Grosse Ache, as the main stream is called, after it has received the two above-noticed tributaries. At Erpfendorf the traveller turns aside into the lateral glen called *Ausserwald*. The head of this branch of the Grosse Ache is not divided by any apparent ridge from the glen of the *Strubach*, a tributary of the Saale, and on the watershed, 2,518 ft. above the sea, stands the village of

Waidring (Inn: Post, good and clean). From hence to Reichenhall the scenery is so remarkable that travellers do well to traverse it on foot. The position of the village is singular. To the N. is the range of the *Hohe Platte*, whose upper part presents a range of absolutely vertical rocks sustaining an upper terrace of alpine pastures. Between this and the more westerly summit of the *Fellhorn* (5,737′), a path mounts to the upper level of the mountain, where the rocks abound in fossils. There is an extensive view, especially on the N. side.

A very interesting excursion may be made from Waidring, through the upper glen of the Strubach, to the Piller See. The way lies through an extremely deep and narrow defile, whence the name Strub, given to holes and clefts where the surface has been rounded and smoothed by the action of water. Where the glen at length opens a little, the traveller passes the very ancient chapel of St. Adolar, with two fresco paintings, attributed by local tradition to Leonardo da Vinci, who is said to have passed some time in a neighbouring monastery. Hard by is a little inn commanding a fine view of this sequestered glen. A little farther, about 1½ hr. from Waidring, is the *Piller See*, a sheet of blue water, rather more than 1 m. long, and about 500 yards broad, at the W. base of the group of high peaks called Lofer Steinberg. It is apparent that the torrent, in cutting through the defile, has lowered the level of the lake, which once extended a considerable distance farther to S. At the southern end of the present restricted lake basin is *St. Ulrich*, the only village. From hence the ascent of the *Flachhorn* (8,246′), one of the highest of the Lofer Alps, is best effected. As mentioned further (Rte. F), the designation Pillersee, or Pillerseethal, is given in a rather vague way sometimes to the adjoining glen of the Pramathal, sometimes to the entire tract between the Kitzbühler Ache and the frontier of Salzburg. A path leads from St. Ulrich, by St Jacob zu Haus (Rte. F), to Fieberbrunn, and a pedestrian going from St. Johann to Waidring might with advantage take this route, which involves but a slight detour.

The road from Waidring to Lofer descends along the rt. bank of the Strubach through an extremely picturesque glen. The limestone strata hereabouts are nearly horizontal, and the precipitous walls and pinnacles of rock often simulate courses of regular masonry. In the narrowest part of the defile such are really seen, in the remains of walls, and a gateway that formerly closed this frontier post between Tyrol and Salzburg, called *Pass Strub*. Thrice in 1805, and again in 1809, the passage of the Bavarians and French was here resisted with desperate valour by the Tyrolese peasantry. A short way beyond the pass, and about 5 m. from Waidring, the road enters the valley of the Saale close to the beautifully situated village of

Lofer (Inn: Löwe, good), 2,055 ft. above the sea. As seen from the N. this appears to be completely enclosed with an amphitheatre of high peaks, forming collectively the group called *Lofer Steinberg*, of which the *Breithorn* (8,171′), *Flachhorn* (8,246′), *Mitterhorn* (8,326′), and *Ochsenhörner* (8,266′), are the culminating points. Not being a post-station, this place is little frequented by strangers; yet it offers many inducements to the lover of fine scenery and the naturalist, who find comfortable quarters at the village inn. Among other rare plants, *Campanula alpina* has been found on the neighbouring mountains. The road to Saalfelden and Zell-am-See is described in § 45, Rte. C.

The road to Salzburg, which within a distance of 30 m. has passed through portions of four distinct drainage basins, now follows the course of the Saale, until this escapes from the mountains into the plain near Salzburg. Keeping to the l. bank of the torrent, which frets and foams amid huge blocks fallen from the surrounding mountains, the road, keeping nearly due N., enters another defile called the *Knie Pass*, once fortified by an Archbishop of Salzburg, to defend his territory against the Tyrolese; and on issuing from it reaches, at 1,883 ft. above the sea, the village of

Unken (Inns: Post; and another near the mineral spring of Oberrain outside the village). The position is not much inferior to that of Lofer, and several interesting excursions may be made in the neighbourhood.

The *Sontagshorn* (6,438'), lying NNW. of the village, commands a remarkable view, in which the towns of Reichenhall and Salzburg, and the broad basin of the Chiemsee (Rte. B), are conspicuous on one side; while, on the other, several of the snowy summits of the central chain are seen in the openings between the nearer limestone peaks. Along with the ascent of the Sontagshorn an active walker will do well to combine a visit to the upper part of the *Unkenthal*, where the stream, that descends through that glen to join the Saale, has cut one of the most remarkable of the singular clefts that abound in this district. It is called *Schwarzenbacher Klamm*. Those who sleep in a *sennhütte*, in order to reach the summit of the Sontagshorn at an early hour, may lengthen the detour by taking in their way, from the summit to the Klamm, the curious Staubbach waterfall at the head of one branch of the valley of the Weiss Traun (Rte. E).

Between Unken and Salzburg the high-road traverses a strip of Bavarian territory, small in extent, but remarkable for its natural attractions, which are further noticed in the next . Travellers who have occasion to pass the Bavarian frontier, should recollect that Austrian paper-money is taken in the neighbouring states only at a considerable reduction on the nominal value, and is absolutely refused at Bavarian railway stations. Those who intend going by this road direct to Salzburg should cause their luggage to be plumbed before leaving Austrian territory, so as to avoid further trouble on returning to it, some 10 m. farther on.

Before reaching the boundary-line the road traverses the *Stein Pass*—the third defile within a distance of 8 m. —fortified, like the others, during the Thirty Years' War, by the then Archbishop of Salzburg. Beyond this is the frontier hamlet of *Meleck*, with a little inn commanding a fine view of the mountains surrounding Lofer. Here the road leaves the Saale for a short time, ascends nearly due N., and traverses two inconsiderable ridges before it joins the road from Traunstein to Reichenhall, a little way W. of the *Thum See* (1,760'), a little lake or tarn lying at the base of the precipices of the *Staufen* (5,950'). The wooden pipes, conveying the saturated brine from the salt springs to Rosenheim, will attract the traveller's attention as he approaches *Reichenhall*, which place, with its neighbourhood, is described in § 45, Rte. A.

Four trains start daily by the branch railway which takes travellers in 1 hr. to Salzburg; but many prefer to follow the road. This passes the ancient Augustinian monastery of St. Zeno, suppressed in 1803, and now a school. In the church, dating from the 12th century, the sculptured portal, the monuments of many old families of Bavaria, and the cloisters deserve notice.

Throughout the way to Salzburg the most conspicuous object is the massive range of the *Untersberg* (6,509'). The part played by this mountain in the mythic tales and popular traditions of South Germany is well known to most readers. Sometimes it is Charlemagne, sometimes Frederick of Hohenstaufen, and sometimes the last great emperor, Charles V., who is held in a magic sleep within the caverns with which the mountain abounds, and who is to issue forth when Germany is restored to unity and glory.

SALZBURG (Hotels: Europa, large new house opposite the rly. station, with a fine view; Nelböck, near the rly. station, excellent; H. d'Autriche, by the river, handsome new house; Goldenes Schiff; Erzherzog Karl; besides the following second-rate but not bad inns—Krone; Mohr; Hirsch; Tieger) was long the capital of one of the ecclesiastical sovereignties which were abolished during the Napoleonic wars. The Archbishop preserves the title of prince, but retains no temporal jurisdiction. This small city is famed for the beauty and picturesqueness of its position. It must, perhaps, yield the palm to Edinburgh and Verona, but, in these respects, deserves the first rank in Germany. Just where the Salza issues forth from the Alps into the plain country, and bends to NW., its channel is hemmed in between two steep isolated masses of rock. The *Mönchberg*, on the l. bank, is a crescent-shaped ridge with the concave side turned towards the river; on the rt. bank, opposite the SE. end of the Mönchberg, rises the *Kapuzinerberg*.

The principal portion of the city lies in the space enclosed between the Mönchberg and the Salza. A narrow passage at the SE. end allows communication between the suburb, called Nonnthal, and the city, through the Cajetan Thor. At the NW. end the rocks approach so close to the river, that the space for the road entering through the Klausen Thor has been partly obtained by piers of masonry along the river's edge, partly by cutting away the rock. The highest part of the Mönchberg—582 ft. above the Salza, or 1,970 ft. above the sea—is the so-called Schlossberg, whereon stands the citadel of Hohen Salzburg. This, as well as the opposite mass of the Kapuzinerberg, consists of limestone (of cretaceous age?), but the NW. part of the Mönchberg is formed of the tertiary conglomerate called *Nagelfluh*. The old houses in the Gstätten Gasse and adjoining streets are not only built up against the vertical faces of the latter rock, but have cellars, and even dwelling-rooms, excavated into it. The city has suffered severely from the slow but certain action of the weather on this conglomerate, the same rock which caused the famous catastrophe of the Rossberg, in Switzerland. Four serious landslips are recorded: that of 1669 is said to have overwhelmed a convent, the church of St. Mark, and 13 houses, burying not less than 300 of the inhabitants.

We shall briefly note the objects most worthy of attention, premising that none approach in interest the admirable views from the two hills overlooking the city.

The Dom, or cathedral, standing in the principal square of the city, is built in imitation of St. Peter's at Rome, and contains many indifferent pictures. The cupola was destroyed by fire in 1859. On one side is the Palace, or Residenz Schloss; and, on the other, the Neubau, with a permanent exhibition of art-works. In the centre of the Platz is a colossal fountain—Hofbrunnen—deserving a moment's notice.

In the adjoining Mozart's Platz is a full-length statue of the illustrious composer, who was born in the street behind the Drey Alliirten hotel: the house, just opposite the hotel, is marked by a marble tablet.

In the Franciscaner-Kirche, built in the old German style, a remarkable musical instrument, invented by one of the monks, is usually to be heard between 10 and 11 in the forenoon.

The church and cemetery of St. Peter are the most interesting ecclesiastical buildings. Three chapels, one of them hewn into the rock, date from the earliest Christian period. The cemetery, partly enclosed by the rocks of the Schlossberg, is very impressive. Among the monuments are those of Mich. Haydn, of Mozart's sister (Baroness Sonnenburg), and of the Chevalier Neukomm. The library of the adjoining monastery is rich in early printed books.

Near the cavalry barracks is the Sommner-Reitschule (entrance 10 kr.), an amphitheatre, with three rows of seats, excavated in the rock. Near at hand is the Neu-Thor, a gateway, with a tunnel 200 ft. long, carried under the lowest and narrowest part of the Mönchberg, giving an entrance to the town from the SW.

The Museum (open from 10 to 4— entrance 20 kr.), containing antiquities, coins, and collections illustrative of the natural history of the province, along with a pretty good library, is on the Quay, near the Ursuline convent.

If the traveller next turns his attention to the rt. bank of the Salza, he sees to his rt., above the bridge, in the centre of the city, only a line of old houses, forming a long narrow street at the foot of the Kapuzinerberg; but, on the other hand, a considerable suburb—the Linzer Vorstadt—extends northward within the line of the ancient fortifications. At the corner of the small Platz, opposite the bridge, is the house once inhabited by Paracelsus; and following the Linzer Gasse, in a line with the bridge, the traveller soon finds on his rt. hand the church and cemetery of St. Sebastian, where rest the remains of the same remarkable man. The Archbishop's palace, called Mirabell, is an extensive pile, with a large garden commanding fine views.

Outside the Linzer Vorstadt is the railway station, whence trains start in one direction for Linz and Vienna, and on the opposite side—crossing the Salza —for Rosenheim and Munich.

No stranger visiting Salzburg should fail to ascend either the Mönchberg or the Kapuzinerberg, if he be not tempted to enjoy the admirable views from both. Though not quite so high as its rival, the Mönchberg offers more variety. The hill, which presents equally steep faces nearly in every direction, forms a grand natural terrace, accessible by a steep carriage road, and by various paths, and planted with trees to such an extent as to limit the view, except from certain favourable points. There are several cafés and restaurants, chiefly frequented on Sundays and holidays. The shortest way to reach the Mönchberg from the middle of the city is by a long flight of 283 steps, near the cavalry barrack, but it appears to be a better course to leave the town by the Klausen Thor and ascend the hill by its northern end, following the road which ascends from the Augustiner Kirche, in the Mülln suburb. The view over the plain and the course of the Salza below the city soon begins to widen out. The Haunsberg and the church of Maria Plain are conspicuous on the rt. bank of the Salza. Bearing to the rt., after he has gained the summit of the hill, the traveller may easily gain a point whence he looks down upon the city and the course of the Salza, and beyond the nearer eminences, to the outliers of the Alps, that are here near at hand. Due E. is the Gaisberg (4,399'), but the higher peaks to the S. are not yet visible.

The point whence the traveller gains this view of the city is very near the spot where the destructive fall of the cliff occurred in 1669. To prevent as far as possible a similar misfortune, men are annually let down by ropes, whose business it is to remove loose portions of the rocks before the cracks extend to a formidable depth.

To gain the view to S and SW., hitherto concealed by the higher part of the ridge, it is now expedient to cross the narrow neck by which the two portions of the ridge are united, above the tunnel of the Neu-Thor, when, bearing to rt., a new and very different prospect is gradually opened out. First, the valley of the Saale, with the Staufen (5,950′), and other higher summits in the distance, then the huge mass of the Untersberg (6,509′) comes into view. Its outline has been often compared to that of the Egyptian sphynx. The plain between the Saale and the Salza, richly planted, and bright with cheerful villages and the castles of wealthy proprietors, fills the middle distance, with the Hohe Göll (8,266′) and the group of the Tünnengebirge (7,965′) in the background. To the SE. the *Gennerhorn* (5,736′) crowns the outer range of the Alps. The stranger may return to the town by the citadel. To visit this a stranger requires a special permission from the commandant. The view from the Glockenthurm comprehends at once all the separate views that are gained from the different points above mentioned. To reach the *Kapuzinerberg*, the traveller must cross the central bridge and follow the Linzer Gasse for a short distance. A large cross on his rt. hand marks the gateway leading to the long flight of steps that give access to the Capuchin convent. It is usual to pay a trifle for permission to enter here. Ladies are not allowed access to the garden, which commands an admirable view of the city and the river. Leaving the convent on one side, a winding path, for the most part well shaded, leads up to the Franciscus Schlössl, standing on the summit of the hill, 2,200 ft. above the sea. Passing through the building, the visitor reaches a bastioned terrace on the opposite side. The finest views are gained from the two corner towers of the bastion, and these perhaps surpass any single view from the Mönchberg. Ladies who are not admitted to the convent garden may enjoy nearly the same view of the town by turning to l. in the descent from the summit, and reaching a spot called the Stadt Platz, where the trees have been cleared away. The effect is admirable. The botanist will be surprised to find on these low hills a considerable number of plants usually considered characteristic of the Subalpine region, along with a few southern species such as the cyclamen and the orange lily.

Of the shorter excursions from Salzburg the following may be enumerated:—

1. The castle and park of *Aigen*, belonging to Prince Schwarzenberg, commanding fine views; charge for carriage, halting 1 hr. and returning, with one horse, 1½ flor.; with 2 horses, 2 fl. 40 kr.

2. About 1 hr. beyond Aigen is the castle of St. Jacob, commanding a much more extensive view. The castle is, or was, inhabited by a priest who received strangers, expecting payment.

3. The *Gaisberg* (4,399′), easily reached in 3 hrs. from Aigen, commands a view of the neighbouring region which is inferior only to that from the Schafberg (§ 46, Rte. A); 8 lakes are seen from the summit. One hour lower down is a rustic inn, where food and lodging may be had.

4. The castle of *Hellbrunn*, with picturesque gardens and a park much frequented on Sunday afternoons, when omnibuses ply every half hour, lies on the road to Hallein, on the l. bank of the Salza. By a footpath and a ferry over the river the pedestrian may combine this with the excursion to Aigen.

5. SW. of the city, less than 1 m. distant, is the castle of Leopolds Kron, belonging to king Louis of Bavaria. About 3 m. farther is *Glaneck*, with an old castle, and near it the Fürstenbrunn, a spring famous for its abundant flow of delicious water, which forms not far from its source a pretty waterfall. Thence the ascent of the Untersberg is sometimes made, see § 45, Rte. A.

6. On a hill about 3 m. N of the city, on the rt. bank of the Salza, is the church of *Maria Plain*, conspicuous from a distance, and commanding an excellent view. From hence it is pretty evident that, at no distant period, a considerable tract of low marshy country round the city was a lake containing numerous islands, now rising as hills from the plain. Charge for a carriage to Maria Plain, waiting 1 hr. and re-

turning, with one horse, 2 fl. 20 kr.; with two horses, 3 fl.

There are numerous public conveyances plying daily from Salzburg to the neighbouring towns and villages at extremely low fares; thus the fare in the stellwagen to Reichenhall is only ½ fl., or one shilling. Diligences run in summer to Ischl and Gastein; and a mallepost (very slow) to Bruck-and-er-Mur (see § 53, Rte. E).

The city contains several good shops, better supplied than in any town equally near to the German Alps. If he may judge by his personal experience, the writer would advise travellers who carry English circular notes not to exchange them here, unless they be prepared to lose as much as 4 or 5 per cent. on the amount of each note.

Route B.
MUNICH TO SALZBURG, BY RAILWAY.

	Bavarian miles	Eng. miles
Rosenheim	9	41½
Stephanskirchen	¾	3½
Endorf	1¼	5¾
Prien	1	4½
Bernau	¾	3½
Uebersee	1	4½
Bergen	1	4½
Traunstein	¾	3½
Teisendorf	1½	7
Freylassing	1¼	5¾
Salzburg	¾	3½
	19	87½

Although the line of rly. between Rosenheim and Salzburg merely skirts the northern base of the Alps, it seems desirable to give a brief notice of the district through which it passes.

For the rly. from Munich to Rosenheim, where the branch leading to Innsbruck joins the line to Salzburg and Vienna, see § 43, Rte. B.

After crossing the Inn the rly. runs eastward to *Stephanskirchen*, on the W. shore of the *Simmsee*, and thence NE. to *Endorf* station. A short way beyond the latter place it approaches near to the Chiemsee, the largest lake of the German Alps. To avoid this, the rly. turns abruptly, and is carried nearly due S. for at least 7 m., passing the *Prien* station, and resuming its easterly direction at *Bernau*.

[The traveller who wishes to reach the Chiemsee from Innsbruck, without making the circuit in a rly. carriage by Rosenheim, may take an agreeable walk through the *Prienthal*. The glen is traversed by a road, but is better suited for a pedestrian.

Leaving the Innthal at *Ebs*, about 5 m. NE. of Kufstein, the track ascends gently somewhat E. of N. to a low pass, about 3,000 ft. above the sea, which marks the boundary of Bavaria. The stream which descends from the pass is the *Prien*. The highest village is *Sacharang*, a very ancient place, mentioned in the chronicles of the 8th century. As the traveller descends the glen the scenery improves. He passes the picturesque castle of *Hohenaschau*, seat of the counts of that name, standing at the W. end of the *Kampen* (5,333′), a range that extends hence to the Grosse Ache. Here the traveller issues from the mountains. One road to rt. goes direct to the Bernau station, and another longer road follows the stream to the village of Prien, near to which it falls into the lake.]

The Chiemsee measures about 12 m. from E to W., and nearly 9 m. from N. to S., but it is apparent from the extent of low land surrounding it, and the number of small lakes lying near its N. and NW. shores, that it has, at no distant date, spread much farther than it now does.

The shores are low, but the range of the Kampen and Hochgern, rising only a few miles to S., and other more distant summits to SE., make a fine background to the views of the lake. It contains several islands. The largest of these, called *Herrenwörth*, is more than 1 m. in length and breadth. A stately building, once a Benedictine monastery, but now private property, is seen from a distance. It stands in a park of some extent, and the neighbouring inn is frequented by summer visitors. The much smaller island of *Frauenwörth* contains, along with the nunnery, which gives it its name, a group of fishermen's houses, and an inn which is a favourite summer resort of Bavarian painters. The old post-road from Munich to Salzburg passed by the N. side of the lake, touching the shore at *Seebruck*, where the river *Alz* flows out to join the Inn only a few miles above the confluence of the Salza. About 4 m. from Seebruck is *Seeon*, an ancient convent, standing on a little lake. This has been converted into a bathing establishment, said to be well-managed and comfortable. The inns in the villages surrounding the Chiemsee are, as a general rule, tolerably good, and are frequented by summer visitors from the Bavarian towns.

The rly. is carried due E. from Bernau, at some distance from the swampy shore, and after passing *Uebersee*, crosses the Grosse Ache, the chief feeder of the lake. The main branch of this stream is the Kitzbühler Ache, and the traveller who would follow one of the great highways of the middle ages, now fallen into disuse, may enter Tyrol by following the stream to Kitzbühel, and thence to the Thurn Pass (Rte. C). Near the rt. bank of the Ache is seen the old castle of *Marquartstein*, now the property of Baron Tautphöus, at the foot of the *Hochgern* (5,681') which commands the finest view of this neighbourhood. The *Bergen* station is some way from the village of that name which is ¼ hr. from the Baths of Adelholzen (Rte. I.) Here the rly. turns NNE. to the little town of

Traunstein (Inns: Hirsch, very good;

Post; Weisses Bräuhaus), rebuilt since 1851, when the larger part was destroyed by fire. It is well situated on the banks of the Baierische Traun (see Rte. E), just 1,930 ft. above the sea. Henceforth the rly. follows a tolerably direct course a little S. of E., through an undulating wooded tract, passing the ruined castle of Raschenberg before reaching *Teisendorf* (Inn: Post). The next station of *Freylassing* is on the W. side of the Salza, where the luggage of travellers entering Bavaria from Salzburg is sometimes examined. A branch rly. is now open from Freylassing to Reichenhall (§ 45, Rte. A). Following the main line, the traveller very soon reaches *Salzburg* (described in last Rte.).

ROUTE C.

WÖRGL TO MITTERSILL, IN PINZGAU, BY KITZBÜHEL.

	Eng. miles
Hopfgarten	5
Brixen	6
Kitzbühel	9
Thurn Pass	13
Mittersill	5
	38½

This is the most direct road for those who travel by carriage from the valley of the Inn to Gastein, or to Styria, through the Pinzgau. On the whole it may be considered as interesting as that by the Zillerthal and the Gerlos Pass. There are no post horses, but carriages may be hired at Wörgl and (usually) at Mittersill. The road from

Wörgl to Kitzbühel ascends the Brixenthal by the l. bank of the stream, on the opposite side to the Kaiserstrasse, described in Rte. A. A stellwagen plies on this road as far as *Hopfgarten* (Inn: Paulwirth), whence travellers usually make the ascent of the *Hohe Salve* (5,993'), an isolated mountain commanding an extensive alpine panorama. It is commonly called the Rigi of Tyrol, but the name is inappropriate, as the lakes which are the characteristic features of the Rigi panorama are here wanting. Horses (fare 4 fl., for going and returning) and tragsessel, or chaises à porteurs (costing 12 fl.), are found at the inn. The way is so well marked that a guide is not necessary. In 3 hrs. the little inn and chapel, standing on the very summit, are easily reached. Rough quarters may be had here for the night. The main features in the view are the snowy peaks of the central range from the Brenner to the Gross Glockner, and the limestone peaks that culminate in the Uebergossene Alp (§ 45). In 1863 no less than 1,052 visitors wrote their names in the strangers' book at the summit.

A short way above Hopfgarten is *Haslau* where two considerable lateral glens— Kelchsauerthal from S., and Winnacherthal from SSE.—pour their torrents into the main stream of the Brixenthal. Following the latter, between slopes formed of a crumbling argillaceous schist, the traveller reaches

Brixen (Inn, tolerably good), not to be confounded with the important place of that name in the valley of the Eisack. Near at hand is the *Maria-Luisenbad*, a somewhat frequented mineral spring. Soon after passing Brixen the road reaches the watershed between the Inn and the Kitzbühler Ache, only about 2,400 feet above the sea-level, and a little way farther, the rather large village of *Kirchberg*. The stream running through the village is not, however, the main branch of the Ache. It is the tributary torrent which, under the name *Rheinach*, flows into the Kitzbühler Ache just above the village of St. Johann (Rte. A). It originates in two alpine torrents, flowing respectively from the E. and W. flanks of the *Rettenstein* (7,750'), one of the highest summits on the N. side of the upper Pinzgau. These torrents, uniting at the alpine village of *Aschau*, run due N. through a glen, which above Kirchberg bears the name *Spertenthal*. Below that place the stream bears a little E. of N., and finally turns nearly due E. to St. Johann (Rte. A.). A path through the Spertenthal leads to Mühlbach in Pinzgau over a pass called *Stange* (5.701').

The road passes the ruins of the very ancient castle of Löwenburg, and skirts the Schwarzsee, which lies at the S. end of a district called *Bühelach*, occupying the space between the Rheinach and the Kitzbühel Ache, and said to present an aspect as though mountains had fallen down and lay in ruins partly covered over by verdure,

Kitzbühel (with a good inn, landlady, Frau Tiefenbrunner, odd-tempered) is a neat little town, 2,480 ft. above the sea, of importance in the middle ages, when the valley from the Chiem See to the Thurn Pass was a frequented route into Italy. The *Kitzbühlerhorn* (6,544') and several other high summits overlook the town, but the clay slate of which they are formed does not weather into such striking forms as those of the limestone mountains about Lofer (Rte. A). The neighbourhood has yielded several rare, and some new, species of plants to the persevering researches of the late Herr Traunsteiner. Among these may be noted *Carlina nebrodensis* (on the Jochberger Alp, and the Sintersbachgraben), *Pedicularis asplenifolia* (on the Geisstein), *Andromeda polifolia*, *Orchis Traunsteineri*, *Malaxis paludosa*, and *M. monophyllos* (all four near the Schwarzsee) *Carex pulicaris*, *C. pauciflora*, *C. tetrastachya*, *C. Gaudiniana*, and *C. microstachya*. The little town has produced several other good local naturalists, and rich mineralogical and entomological collections were, and probably still are, to be found here.

The view from the *Kitzbühlerhorn*

is held by Dr. Ruthner to be the finest in this district, but can scarcely equal that from the Geisstein. Those who wish to enjoy it in perfection sleep at the Dratalp, only 1 hr. below the summit. As it is equally accessible by the N. side, towards St. Johann, it may be taken by the mountaineer on his way from one place to the other.

The road to Mittersill crosses to the rt. bank of the Ache, but after reaching *Aurach* (Inn: bei Joseph Filzer), a small village with an ancient church, it returns to the W. side of the stream. The next village, about 6 m. from Kitzbühel, is

Jochberg (two Inns: the best bei Wagstetten). The same name is given also to the upper part of the valley from hence to the Thurn Pass. One of the inns here was long kept by Oppacher, one of the heroes of the Tyrolese struggle against the invasions of 1805 and 1809. In the latter year he held with a small force the Strub Pass, near Lofer, for 9 hrs. against 10,000 French troops, and later in the same year stormed the enemy's position at the Knie Pass (Rte. A), taking a number of prisoners not much inferior to that of his own men.

The ascent through the upper valley is gentle, and in 7 m. from the village the traveller reaches the summit of the *Thurn Pass* (4,371′). The old road passed by a group of houses called Spital, from a hospice which formerly stood there. A decent inn stands by the road at the summit of the present passage, and commands a fine view in both directions. [The mountaineer may enjoy a very much grander prospect, probably the finest in this district, by ascending the *Geisstein* (7,747′) lying E. of the pass, and forming the corner stone of the Pinzgau, the valley of Kitzbühel and the Glemmthal (Rte. G). The range of the Hohe Tauern, from the Glockner to the Dreiherrnspitz, is here seen to full perfection, and the remainder of the panorama is scarcely less interesting. The traveller should sleep at Jochberg, and take on his way the

Sintersbachfall, 200 ft. in height, decidedly the finest waterfall of this district. A guide is necessary.]

The road descends rapidly on the Pinzgau side of the Thurn Pass, turning eastward as it approaches the level of the valley, and in 5 m. reaches *Mittersill* (Inns: bei Grundtner, in the market-place, best; Bräu Rupp, large house, not well managed), described in § 50, Rte. A.

ROUTE D.

HOPFGARTEN TO WALD, IN PINZGAU.

The mountaineer who, having visited the Hohe Salve from Hopfgarten, wishes to reach the head of the Pinzgau, need not make the detour by Kitzbühel, described in the last Rte., as there are two alpine glens which offer a more direct way, one of which serves equally well, if his object be to visit the Zillerthal.

1. By the *Winnacherthal*. This glen, also called Windau, opens into the Brixenthal at Haslau, a little above Hopfgarten. It affords the most direct way to the head of the Pinzgau, as the track over the pass at the head of the glen descends close to Wald, 1½ hr. from Kriml, and about 17 m. above Mittersill.

2. By the *Kelchsauerthal*. This is a less direct way to the Pinzgau, but is said to offer finer scenery than the last. The opening of the glen is about half-

way between Hopfgarten and Haslau. By the ruins of the castle of *Engelsberg* is a very ancient chapel, called Elsbethenkirchlein, concerning which a curious legend is still told by the country people. About 2 hrs. from the road is the village of *Kelchsau*, which gives its name to the glen. Less than 1 hr. farther the valley divides. The shorter l. hand branch, called Kurze Grund, leads to the *Salza Joch* (6,533'). By a detour, with the aid of a guide, the traveller may reach a little alpine lake at the base of the lofty *Geyerkopf*. From this the Salza is said to originate, and it is easy to follow its course down to a second lake at a lower level. The direct way over the pass is free from difficulty, and the traveller descends to Ronach, the highest hamlet in Pinzgau, 2 hrs. above Wald.

The other longer branch of the Kelchsauerthal, called Lange Grund, is best suited for those who wish to reach the Zillerthal from Hopfgarten. The pass at the head of the glen lies between the *Thorhelm* (8,518') and the *Stulkorkopf*, and leads to the track of the Gerlos Pass a little above the village of that name.

Dr. Ruthner has given, in the second vol. of the proceedings of the Austrian Alpine Club, an account of the ascent of the Thorhelm effected by him, in company with Herr Unterreiner, the head forester at Gerlos. The summit, which by a strange error is placed on Scheda's map S. of Gerlos, is reached in 4 hrs. from the inn at that village (§ 50, Rte. A) without any serious difficulty. It commands a fine view of the Zillerthal Alps. It is likely that an active mountaineer may take the summit on his way between Gerlos and Hopfgarten.

The measurements given in several German works for the Stulkorkopf—9,085 ft., and for the Geyerkopf, 9,062 ft.—are certainly erroneous. Dr. Ruthner satisfied himself that the Thorhelm is the highest summit of the group. The next in height is probably the neighbouring peak of the *Katzenkopf* (8,311').

The way from Zell, in Zillerthal, to the Pinzgau is described in § 50, Rte. A.

Further information as to both the above-mentioned passes is much desired.

Route E.

TRAUNSTEIN TO KITZBÜHEL.

The traveller, who wishes to avoid beaten tracks, may choose between two or three different routes, leading to the central range of the Noric Alps, through the mountains lying S. of the Chiem See, on the frontier of Bavaria, and may enjoy on the way some pleasing, and even grand, scenery.

1. *By the valley of the Grosse Ache.* As mentioned in Rte. B, the Grosse Ache, which unites the principal streams from the Alps surrounding Kitzbühel, issues from the mountains a little way S. of the Uebersee station on the rly. between Traunstein and Rosenheim. In a direct line, the distance from Uebersee to Kitzbühel is scarcely 30 m.; but by the road, following the windings of the valley, it is counted as 16 stunden or 48 m. In its way from the upper valley to the level of the Chiem See the Ache flows through an opening or cleft between the Kampen (5,333') and the Hochgern (5,681'). In this cleft stands the castle of Marquartstein, noticed in

Rte. B. The scenery is wilder and more striking than might be expected amidst mountains of such moderate height. The track is very ancient, and was frequented in the middle ages, though now rarely used except by the villagers. At *Unter Wessen*, the first village, a country road mounts through the lateral glen to S., and joins the route by Reit im Winkel, described below. The main road keeps to the l. bank of the Grosse Ache towards SW., till after passing the village of Schleching, it returns to the rt. bank, and enters the defile of *Klobenstein*. This is one of the most remarkable of the numerous defiles which abound amid the limestone mountains of this district, and may bear comparison with many of those that have attained celebrity. The narrow road, which is passable only for the smallest and lightest country carriages, mounts gradually between the walls of limestone rock, until the roar of the torrent in the depths below is almost lost to the ear. At the highest point attained is the chapel of Klobenstein, and the road, having now entered Tyrol, descends until it again has reached the level of the valley which opens out in the green basin, wherein stands, 1,867 ft. above the sea, the pretty village of
Kössen (fair country inn). This retired place, though scarcely known even by name to strangers, may afford pleasant head-quarters for many interesting excursions. It is a centre towards which converge three lateral glens, besides the main valley, traversed by the road to Kitzbühel. That of the Weisse Lofer is noticed below.

In the opposite direction, from a little S. of W., a stream descends from the pretty *Walch See*. Beside it is the village of *Walchsee* (2,084′), 2½ hrs. from Kössen. A very slight ascent is needed to cross the watershed, between the lake and the *Jenbach*, which flows westward towards the Inn, and enters the Innthal near Ebs, 2½ hrs. from Walchsee, whence Kufstein is reached in 2 hrs. more. See § 43, Rte. B.

From Walchsee, and many other places near Kössen, the most conspicuous objects in all the mountain views are the rugged peaks of the Kaisergebirge, already noticed in Rte. A. Those who would make a nearer acquaintance with this group should explore the deep glen which lies SSW. from Kössen, and joins its torrent to that from the Walch See, near a large brewery, formerly the seat of a noble family, with a chapel, said to contain curious paintings. The lower part of the glen is a cleft through eocene rocks, containing seams of lignite, whence it is called *Kohlnthal*. The glen opens at Schwend, and the upper part is thenceforward called Kaiserthal, being the name also given to the glen, which on the opposite side descends towards Kufstein. On approaching the higher peaks of the range, the track turns SE., and mounts to the village of *Kirchdorf*, whence it descends to St. Johann. The *Eastern Kaiserthal* runs deep into the heart of the Kaisergebirge, which is divided into two groups by this and the other glen—*Western Kaiserthal*. The lower northern group, called *Hinter Kaiser*, little exceeds 6.000 ft. in height. The southern group, called *Vorder Kaiser*, or *Wild Kaiser*, includes several rugged peaks, of which the highest is the *Scheffauer Kaiser* (7,611′). This is best ascended from the S. side. From the N. it is very difficult of access. An easy pass leads from the head of the E. Kaiserthal to Kufstein through the W. Kaiserthal.

The neighbourhood of Kössen is known to geologists by the beds called *Kössener Schicten*, forming part of the Rhætic Group.

Resuming the road along the main valley of the Grosse Ache, which for 3 hrs. above Kössen is locally known as the Kössenthal—a narrow wooded glen with scarcely any trace of inhabitants—the traveller reaches Erpfendorf (1,978′), where the Kaiserstrasse, described in Rte. A, turns aside from the valley of the Grosse Ache towards Waidring. 5 m. further is St. Johann (Rte. A). Here the road to Wörgl leaves the main valley to ascend along the Rheinach to Elmau, but there is a good country road along the main stream, which henceforward bears the name Kitzbühler Ache.

The first village is Oberndorf, commanding a very pleasing view, especially to NW., where the crags of the Kaisergebirge are seen above the undulating heathery tract, called Bühe lach. In 7½ m. from St. Johann the traveller reaches Kitzbühel, for which see Rte. C.

2. *By the Valley of the Baierische Traun.* The chief stream that flows from the Alps into the Bavarian plain between the Grosse Ache and the Saale is the Traun, which, to distinguish it from the more considerable river of the same name in the Salzkammergut, is called *Baierische Traun*. It does not, like the Grosse Ache, penetrate deeply into the mountains: the branches, which unite a little S. of Traunstein to make up the main stream, all rise in the outer range of the Bavarian Alps.

Starting from Traunstein (Rte. B) the traveller follows the road to Reichenhall for about 4 m., as far as Siegsdorf (§ 45. Rte. A). It will be observed that the level of the Traun valley is considerably higher than that of the Grosse Ache. Traunstein is about 200 ft. higher than the Chiem See, and nearly at the same level as St. Johann, where the Ache has penetrated deep into Tyrol; and Siegsdorf, where the Traun enters the plain, is 53 ft. higher than the last-named village, or 2,003 ft. above the sea. Leaving the branch called *Rothe Traun*, along which the road to Reichenhall extends SE., the traveller follows a country road by the rt. bank of the main stream, or Weisse Traun, to Eisenarzt. This village may be reached in about the same time by taking the train from Traunstein to the Bergen station, and going thence to the mineral springs of *Adelholzen*, which enjoy some local repute. There are three springs, varying much in their chemical constitution. The baths are about 2 m. from Siegsdorf, and the like distance from Eisenarzt. The lower part of the course of the Traun is interesting to geologists, who may here trace the junction of the miocene molasse with underlying eocene strata, which at many points abound in fossils. Towards the Chiem See, again, the still newer pliocene beds come into view.

The lover of distant views may, from the Bergen station, ascend the *Hochfellen* (5,356′), a NE. promontory from the range of the Hochgern, which conceals part of the view southward, but leaves a wide panorama to N. and E. Thence it is best to descend to *Ruhpolting* (2,179′), on the l. bank of the Weisse Traun, 3 m. S. of Eisenarzt. Here the valley is narrowed between limestone rocks, but it opens again above the village into an open basin, studded with many scattered farm-houses, where three torrents unite to form the Weisse Traun. The least of these, called *Windbach*, flows from the E. through a broad, nearly level depression, by which the pedestrian may reach Inzell, at the head of the valley of the Rothe Traun (§ 45, Rte. A), in 2 hrs. From SW. flows the *Urschlau*, and by it is the shortest, but not the most interesting, way to Reit im Winkel, mentioned below.

The main stream, now called *Seetraun*, issues from a defile between the *Unternberg* (4,849′) and the *Rauchberg* (5,543′). The latter mountain has been the scene of much ill-rewarded mining enterprise for the last three centuries. Large quantities of lead and zinc have at times been extracted, but the cost of the seventy-two shafts sunk into the mountain, and of other works connected with them, has probably exceeded the value of the produce. Beyond the defile the Seetraun receives the Fischbach from the S., while the track along the main stream turns to SW., which direction is maintained to the head of the valley.

[The traveller should not omit to make a short detour from his route in order to visit a singular waterfall. The glen of the Fischbach is a mere cleft, through which a narrow path has been carried, ascending partly along ledges of the rock, above the rt. bank of the torrent. As he rises higher and higher the gulf on his rt. hand becomes deeper and darker, but the wall of rock to the left continues to rise perpendicular, sometimes almost overhanging. Suddenly the way seems barred, as a waterfall, springing from a ledge overhead, falls to a depth of 600 ft. into the bottom of a chasm. Formerly passengers crept along a plank, thrown across under the arch formed by the falling water. The present more secure path is cut into the rock beneath the fall, protected from falling stones by a wooden roof. The fall, which is called the *Staubbach*, or Staub, is formed by a stream descending from the Sonntagshorn through a lateral cleft into the ravine of the Fischbach. Higher up in this ravine is the *Fischbachfall*, where the main torrent springs over two successive ledges of rock into the dark channel, where, lower down, it is joined by the Staubbach. At the head of the glen the path along the Fischbach traverses an easy pass leading into the upper part of the Unkenthal, and by that way an active walker may reach Unken, or even Lofer, in one day from Traunstein.]

In about 1 hr. from the junction of the Fischbach, the traveller ascending the main valley of the Seetraun reaches the *Forchensee* (2,456'), on whose shore stands a solitary inn called Seehaus. This is the first of a series of small and very picturesque lakes, and is separated from the *Lödensee* by a short gorge. This latter, which is the largest, is succeeded by the *Weitsee*, nearly divided into two separate basins by a projecting point of land. From hence the pedestrian may reach Reit im Winkel by crossing a low pass lying W. of the lake, but the rough cart road turns SSW., till in ½ hr. it enters the head of the glen of the *Weisse Lofer*, and then bearing to the rt., leads to the pretty highland village of

Reit im Winkel (2,196'). Though still in Bavaria, this is but a few hundred yards from the Tyrolese frontier. The neighbourhood is said to be very picturesque, and the bold mass of the Kaisergebirge is conspicuous to SW. The geologist, who may be lodged in one or other of the village inns, will find ample occupation among the fossiliferous rocks of the gorge of the Weisse Lofer, through which that stream descends to join the Grosse Ache at Kössen, where the traveller will join the route from the Chiem See to Kitzbühel, described above.

Route F.

ST. JOHANN TO SAALFELDEN, BY FIEBERBRUNN.

The traveller who has reached St. Johann by the Kaiserstrasse (Rte. A), or who has crossed the Kaisergebirge from Kufstein, and whose aim is eastward, either to Gastein or the Styrian Alps, may take a very direct course to Saalfelden, in the Valley of the Saale, described in the next §.

The Piller See, at the head of the Strubach, which, after flowing from the lake to Waidring, runs along a side of the high-road from that village to Lofer, has been already noticed in Rte. A. The name Pillerseer Ache properly belongs to the upper course of the Strubach as far down as Waidring, but it has also been locally applied to the stream flowing WNW. from Hochfilzen to St Johann, through a glen which affords the most direct passage from the Piller See to St. Johann. That glen is thence often called

Pillerseethal, but the correct name is *Pramathal*. Besides offering the most direct way to Saalfelden, it affords a pleasant detour to the pedestrian bound for Lofer who would take the Piller See on his road.

The lower part of the Pramathal is somewhat monotonous. In 2 hrs. the rough road reaches *Rosenegg* (2,301′), where the iron ore from a neighbouring mine is smelted. Here a country road turns off to l. and crosses the low pass leading to the Piller See. On the watershed between the adjoining valleys stands the hamlet of *St. Jacob im Haus* (2,791′). About 1 m. above Rosenegg is the chief village of the Pramathal, *Fieberbrunn* (Inn: beim Auwirth; good for so remote a place). The village is also called Prama and Pillersee. The first name is taken from a mineral spring, said to have cured Margaret Maultasch of a dangerous fever.

An excursion may be made from Fieberbrunn to the *Wildalpensee*, by a path especially interesting to the geologist, who will traverse in succession several beds referable to the trias, followed by clay-slate, and other strata which have been called Devonian, though their exact age does not seem to have been clearly established. The lake is a dark tarn, 6,660 ft. above the sea, according to Schaubach (but ?), commanding a fine view of the Lofer Alps. The water is of a dark colour, as are the fish, which are said to have a disagreeable flavour.

With a local guide there is no difficulty in going from Fieberbrunn to Kitzbühel, taking on the way the summit of the Kitzbühlerhorn (Rte. C).

About 1 hr. above Fieberbrunn is the opening of the *Schwarzachgraben*, through which the Glemmthal, noticed in next Rte., is easily reached. The most frequented way, partly practicable for rough country vehicles, traverses a pass called *Alte Schanze* (4,335′), descending on the opposite side to Saalbach.

At the extreme head of the Pramathal is *Hochfilzen* (3,177′), rather more than 4 hrs. from St. Johann. The little village, which has a tolerable country inn, stands on a green plateau, whence paths descend in three directions, as the main branch of the true Pillersee glen terminates here, as well as the Pramathal and the *Leogangthal*, through which the traveller descends to Saalfelden. The latter glen marks the line of junction between the secondary limestone of the rugged Lofer Alps and the arenaceous or argillaceous underlying strata. The distinction is apparent in the contrast between the forms of the *Rothhorn* (7,914′) and *Birnhorn* (8,635′) to the N., with the *Durchenkopf* (6,173′), and other rounded summits on the S. side of the glen. Little more than ½ hr. below Hochfilzen the traveller reaches *Griesensee*, a wild tarn, beside which is marked the boundary between Tyrol and Salzburg. Lower down are the remains of a furnace, where nickel was formerly extracted from the ore found in a neighbouring mine, rich in rare minerals; e.g. malachite, strontianite, cœlestine, various ores of cobalt, &c. The chief hamlet is *Leogang*, where there is a tolerable country inn. About 1 hr. farther the traveller reaches the valley of the Saale, and crosses the bridge, which leads in about 1 m. to Saalfelden, described in § 45, Rte. B.

ROUTE G.

KITZBÜHEL TO ZELL-AM-SEE, BY THE GLEMMTHAL.

Parallel with the line of valley through which passes the way described in the last Rte. from St. Johann to Saalfelden, is another, which in a similar manner

MAP OF THE BERCHTESGADEN DISTRICT

serves to connect Kitzbühel with Zell-am-See, and offers a variation on the ordinary road by Mittersill. This valley is the *Glemmthal*, and the Glemmer Ache, which flows eastward through it, is the principal source of the Saale. From the head of the glen to its opening near Zell-am-See is about 6 hrs. walk.

The shortest way to reach the Glemmthal from Kitzbühel is by the *Aurachgraben*, a short glen that descends from SE. to Aurach, on the road to the Thurn Pass (Rte. C). A longer and more interesting path is by the Sintersbach, and a little tarn called *Sternsee*, SE. of the *Gamshag* (6,720'). This is sometimes ascended for the sake of the view, but that from the Geisstein (Rte. C) is much to be preferred. Those who go by the Sternsee to Zell-am-See should sleep at Jochberg rather than at Kitzbühel, as the distance is rather considerable and the pass high (about 6,500'?). About 2 hrs. from the head of the Glemmthal is an inn, whence a path mounts to SW. and leads over the Pihapenkogl to Stuhlfelden, in Pinzgau, a few miles below Mittersill. Henceforward the way down the valley is by a rough country road. Another hour's walk takes the traveller to *Saalbach* (3,269'), the chief village of the valley, where the track from Fieberbrunn, mentioned in last Rte., joins our track. This keeps the l. bank of the Glemmer Ache, or Saale, until, after passing *Viehhofen* (2,721') and approaching near to the opening of the valley, the track crosses to the rt. bank, and soon after, in 3 hrs. from Saalbach, joins the road leading from Lofer to Zell-am-See. A walk of 1 hr. takes the traveller to that village, described in § 45, Rte. B.

SECTION 45.

BERCHTESGADEN DISTRICT.

IF the district described in the last section may be roughly defined, as filling the space between the Ziller and the Saale, that now to be noticed is still more accurately defined as the region enclosed between the Saale and the Salza. On issuing from its parent glen, the Glemmthal, into the open valley N. of the Zeller See, the Saale, here separated from the Salza only by a short nearly level space, turns N., and keeps the same general direction till, near Unken, it leads to NE., and follows that course to the plain near Salzburg. Meanwhile the Salza, after passing close to the S. end of the Zeller See, runs eastward for about 20 m. to St. Johann-im-Pongau, then turns abruptly to N., and keeps the same course for 40 m., till it receives the Saale below Salzburg.

Within this narrow space is enclosed a small group of lofty limestone peaks, surrounding a lake of little extent, but scarcely equalled elsewhere for the grandeur and beauty of its scenery. The Königs See and the basin which is drained into it belong to Bavaria, along with the tract lying between it and Reichenhall; but the western, southern, and eastern slopes of the same mountains lie within Salzburg territory. The small tract thus almost enclosed within the Austrian frontier is justly deemed a jewel in the Bavarian crown. The exquisite scenery has rendered this the favourite resort of successive sovereigns, while the salt mines of Reichenhall and Berchtesgaden have produced a large revenue to the royal exchequer.

Berchtesgaden is the natural centre of

this district, and the point chiefly resorted to by foreigners. The mountaineer will prefer the new inn at the N. end of the Königs See. There are several other places offering attractive headquarters. Of these the most frequented by German visitors is Reichenhall, while Ramsau, resorted to by artists and naturalists, has rougher accommodation, but greater natural advantages.

The mountains immediately surrounding the Königs See do not attain to 9,000 ft., and only two of them, Watzmann (8,988'), and Schönfeldspitz (8,696'), approach that limit. The much higher summit of the Uebergossene Alp (9,643') lies S. of the frontier of Salzburg, and is not seen from the lake.

On the S. side of the high group of limestone mountains is a tract of much less height, being in truth an eastern extension of the zone of triassic and palæozoic rocks, noticed in the introduction to the last section. The comparatively low mountains of this small tract are sometimes called Dientengebirge, from Dienten, the chief village. The rivers which so nearly enclose this district are more correctly written Salzach and Saalach, but the ordinary spelling, Salza and Saale, are here adhered to.

ROUTE A.

MUNICH TO BERCHTESGADEN, BY REICHENHALL.

	Bavar. miles	Eng. miles
Rosenheim (by rly.)	9	41½
Traunstein (by rly.)	6½	30
Inzell (by road)	2¾	11½
Reichenhall	2½	11½
Berchtesgaden	2¾	11
	22⅔	104½

The rly. from Munich to Rosenheim is described in § 43, Rte. B, and that from Rosenheim to Traunstein in § 44, Rte. B.

A branch rly. is now open between Freylassing (§ 44, Rte. B) and Reichenhall, and this is rather the shorter way for travellers from Munich to Berchtesgaden, but the road from Traunstein to Reichenhall is by far the most interesting and agreeable way. A diligence plies, or did ply, twice daily from the rly. station at *Traunstein* to Reichenhall, besides which, carriages are usually in readiness to convey those who prefer to enjoy the scenery. The road keeps to the l. bank of the Traun as far as *Siegsdorf* (§ 44, Rte. E), a little above the junction of the Rothe Traun with the main stream of the Baierische Traun, and after crossing the latter, follows the l. bank of the former stream. The wooden pipes conveying brine from Reichenhall to Traunstein and Rosenheim are frequently seen along the road. This crosses the Rothe Traun about 4 m. from Siegsdorf, and in 3 m. farther attains the head of the valley, where a level green plain occupies the site of a lake that once fed the stream. At the farther end stands the pretty village of

Inzell (2,262'), a post-station with a large inn. [An easy walk over low hilly ground leads thence to Ruhpolting, in the valley of the Weisse Traun (§ 44, Rte. E).] The traveller here finds himself at the portal of the Alps. Looking S. the rocks of the *Falkenstein* (4,265') to the l., and the *Kienberg* to the rt. form the gateway, and the fine peak of the Watzmann, in the back-ground, invites the mountaineer onward. Following the road through the opening, he finds it gradually enlarged between the bases of two higher mountains—the *Staufen* (5,950′) to l., and the *Rauschenberg* (5,543′) to rt.—between which a narrow stream, called Weissbach, descends towards the Saale. The road is carried for some miles at a considerable height above the l. bank, until the road from Lofer, having crossed the same stream, ascends the slope to join our route. A sharp turn to NE. carries the road aside from the junction of the Weissbach with the Saale, and then a rather long descent leads to

Reichenhall (Inns: Krone, very good; Russischer Ho.; Löwenbräu; Hohenstaufen), an ancient town, 1,538 ft. above the sea, almost completely rebuilt since 1834, when the greater part was burned down. Besides the above-named hotels, and several others of inferior rank, there is a large hotel and pension, said to be well managed and reasonable, in the castle of Kirchberg, just out of the town on the l. bank of the Saale, and the still larger bathing establishment of *Achselmannstein*, also near the town, by the road to Salzburg.

A large number of German visitors resort annually to these establishments, chiefly for the sake of the baths of warm concentrated brine, obtained from the neighbouring boiling-houses. Various other curative agents are, however, employed by the patients. Goat's whey as a drink, as well as for baths, the juice of herbs, inhalation of the vapour from the boiling-houses, and that of pulverized medicaments on the system, lately introduced in France and Germany, are all more or less in vogue. Those persons who travel for relaxation are not, however, likely to prefer this place to luxurious Ischl, or even to rustic Berchtesgaden; yet it is not without natural attractions, and those who may be induced to spend some time will find many short walks and drives, as well as longer excursions, to the places noticed in this section, nearly all of which are within a single day's ride or drive. Carriages and light one-horse vehicles are hired on reasonable terms. For the latter the fare to Berchtesgaden, Salzburg, or Inzell, is 4 fl. ; to the Königs See 5 fl. For a 2-horse carriage, from one-third to one-half more,

The name of this place indicates the substance to which it owed its early importance and continued prosperity, as the early dialects of South Germany, like the Greek, replaced the sibilant in the Latin *sal* by an aspirate. Unlike Hall in the Innthal, Hallstadt in the Salzkammergut, Hallein in the Salza valley, and the neighbouring Berchtesgaden, Reichenhall does not furnish salt in the solid shape. The precious mineral is here obtained in the form of a concentrated solution. The numerous brine springs are on the flanks of the *Gruttenberg*, a hill composed of tertiary rock (nagelflue), and are reached by a shaft that penetrates 54 ft. below the surface. The stronger brine is raised in pumps, and then carried in pipes to the boiling-house, while the weaker springs and waste fresh water are carried to the Saale by a large subterranean drain of solid masonry, 1½ m. in length, through which visitors are conducted in a boat. Besides the produce of the springs, the saturated solution obtained from the mine at Berchtesgaden is also conveyed here. Although the evaporating-houses and works connected with them are on a very large scale, the main building being nearly half a mile in length, only a portion of the brine is here converted into salt, the residue being conveyed to Traunstein or to Rosenheim. The machinery for raising the brine, and driving it through the brine conduits (*Soolenleitungen*) to so great a distance, is said to be very ingenious, and to deserve the attention of engineers. This is to be seen at the Hauptbrunnhaus. The chapel in that building contains three windows which are good specimens of modern Bavarian stained glass. The old church of St. Nicholas, in the town, has been recently restored and deserves a visit.

There is also a local museum, and a private collection of fossils belonging to a person named Mack. The neighbourhood is interesting to the geologist, as fossils of many periods, from the lias to the nummulitic limestone inclusive, are found in tolerable abundance.

Of short excursions, one of the most agreeable is that to the chapel of St. Pankraz, W. of the town, near the little Thumsee. The *Högelberg* (2,676'), between the road to Teisendorf and the l. bank of the Saale, is a hilly wooded tract of considerable extent, where the pedestrian may discover for himself pleasant nooks, and occasionally gain fine views, especially from the castle of Rachenluegg.

Visitors often extend their stroll across the Austrian frontier to the villages of *Grossgmein* and *Marzoll*, at the foot of the Untersberg. That mountain, so conspicuous in all the views of this neighbourhood, is easily ascended from the former village. (See excursions from Berchtesgaden.)

The *Staufen* is a ridge whose highest point, called for distinction Hohe Staufen (5,950'), is not very often ascended. Near the summit the geologist will find the so-called Raibl beds with characteristic fossils, but the botanist has little to reward his pains. Most visitors content themselves with climbing the *Zwiesel* (5,757'), or middle summit of the Staufen; 3 hrs. suffice for the ascent, which is very easy, and the way is pointed out by finger-posts.

The road between Reichenhall and Berchtesgaden is very agreeable, especially when travelling to the latter place. To go from one place to the other it is necessary to pass from the valley of the Saale into the confined basin which pours its waters into the Salza through the Albe.

The low and broad *Halthurn Pass* (2,225'), connecting the Untersberg with the Dreysesselkopf (5,892'), is traversed by the road which, during the ascent from Reichenhall, crosses a narrow strip of Austrian territory. As the traveller advances the Hoher Göll (8,266'), the range of the Hagengebirge (7,710'), the Schönfeldspitz (8,696), and the Watzmann (8,988') appear in succession. The road, which is the ancient way by which salt was conveyed from Berchtesgaden, is called the Hallstrasse. On descending slightly from the pass the road follows a petty stream that flows through an open valley to reach *Berchtesgaden* (Inns: all inferior to those at Reichenhall; Watzmann, best situated, fairly well kept, but prices rather high; Leuthaus, or Post, reasonable, fair accommodation; Neuhaus, of less pretensions than the others, but well recommended; a new hotel—H. Wacker [?]—has also been favourably spoken of) is but a large scattered village, 1,987 ft. above the sea, built on undulating ground above the Albe, and commanding charming views of the neighbouring mountains. The late King Max built a shooting lodge just outside of the village, where he usually spent a portion of the summer. It commands a fine view, but is not otherwise remarkable. Josef Grafl is probably the best guide for mountain excursions. His two brothers are also recommended.

A little below the village the Seeache, from the Königs See, and the Ramsauerache, flowing from WSW., unite to form the Albe. On the S. side of that stream, a little below the village, is the *Salzberg*, one of the most important of the numerous salt mines for which this portion of the Alps is famous. Twice a day (between 10½ and 12, and from 4½ to 6) parties of visitors are admitted by tickets (charge, 45 kr.) issued at the head office in the village, and easily procured through any of the innkeepers. Those who seek admission at other times have to pay an extra fee of 2 flor., besides paying for tickets at the above rate. The mine is not so extensive as that of the Dürnberg, near Hallein (Rte. E), but it is less inconvenient to visit, and the fees are lower. The general plan of working is the same in both mines. The salt is more compact, and masses of pure rock-salt are not uncommon. As in most underground expeditions, however, the visitor sees but little. In one of the

ROUTE A.—THE KÖNIGS SEE.

principal galleries he is half stunned by the explosion of a little mortar; and, on payment of an extra fee of 30 kr., he traverses in a boat a subterranean pool of brine. He returns to daylight astride on a sort of wooden horse that runs on an underground railway, and leaves at the office the suit of over-clothing which was supplied at his entering the mine.

Berchtesgaden has long been famous for its carved ware in wood, bone, and stag's horn, of which a large assortment is kept for sale at Kaserer's warehouse. The workmanship is superior to that of the common Swiss and Tyrolese articles, and the prices very reasonable.

Of the numerous excursions that may be made from Berchtesgaden that to the *Königs See* far surpasses the rest in interest. This indeed is the main attraction that leads so many visitors to this district. The distance is about 3½ m., and there are two ways, one by carriage road on the rt. bank of the Seeache; the other, still more pleasing, by a new shady footpath which turns off near the boiling houses, below the village, keeps all the way to the rt. bank of the bright blue-green stream. At the Königs See end a signpost directs strangers to it. The carriage road passes the hamlet of Unterstein, where parties from Berchtesgaden often go to dine at a house kept by the author of an esteemed cookery book.

The road and the path both meet at the little village of *Königs See*. There are here two inns. The old house supplied tolerable beds and indifferent fare; of late a larger inn, providing 40 beds, has been opened, which the mountaineer would be tempted to make his headquarters but that there have been serious complaints of extortion. Beside the new inn is the house of the head boatman, who arranges for the conveyance of visitors. The charges are fixed by tariff at a very moderate rate. For going as far as the Kessel each rower is entitled to 18 kreutzers; to St. Bartholomä, 30 kr.; to the Sallet-Alp, at the head of the lake, 42 kr. In addition to this the charge for a small boat for the day is 16 kr.; for a larger boat, 40 kr.; and for a large barge, 1 fl. Half these prices are charged for half a day. The boats are very rude in appearance, and the rowers are often stout peasant girls.

From the landing-place at Königs See the visitor sees but a small part of the lake. Each stone in the shallow bed is seen as he gradually advances past a little island and rounds the Falkenstein, a bold rock rising out of the water on his right hand. It is only when this is reached that the main reach of the lake opens before him. The effect is unexpectedly grand and impressive. The sheet of dark blue water, about 6 m. in length, and nearly 1 m. in breadth, 1,996 ft. above the sea level, is encompassed by mountains that at most points literally rise from the water's edge, and often in walls of rock so steep as to be inaccessible even to the chamois that abound on the surrounding crags. But the forms of the rocks are wonderfully varied, and many a ravine runs down to the lake, where, as well as on each shelf of the mountains, dark masses of pine contrast with the reddish or pale grey hues of the limestone rocks. If the back-ground presents no objects equal in grandeur to the ranges of snowy Alps seen from several of the Swiss lakes, the nearer views are not, in the writer's opinion, surpassed by any of them.

The first object pointed out on the E. shore is the *Königsbach*, a slender torrent that falls in a cataract. Once or twice in each year the water of this torrent, which has been retained by a dam in the upper valley whence it flows, is used to carry down a quantity of timber from the upper shelf of the mountain. The effect as viewed from the lake is remarkable. A little way farther the rowers halt at a point where it is usual to discharge a pistol, for the sake of the echo that reverberates in a prolonged roll between the opposite rocks. The effect is singular, and should not be missed. The charge for each discharge is 8 kr. The lake is here at its greatest depth, about 660 ft. Near this is a cleft in the eastern shore, called Kuchler Loch, where it appears that a portion of the water of the lake escapes through a subterranean channel. It is believed that this feeds the fine waterfall near Golling, mentioned in Rte. E. On the same side of the lake, a little farther on, it is usual to land at a place called the *Kessel*, where the Kesselbach torrent descends through a narrow ravine. An ascent of 10 min. leads to a little double cascade, not remarkable as a waterfall, except after wet weather, but well deserving a visit for the sake of the striking scenery around, and the charming views gained in redescending. The mountains on the E. side of the lake abound in game, especially chamois and roe-deer, and are the frequent scene of royal hunting parties, for which purpose the late king opened a bridle-track, by which the traveller may mount in 3 hrs. from the Kessel to the *Gotzenalm* (5,527′), then pass southward to the *Regenalm* (5,196′), where there is a royal shooting-box, and descend by the Kaunerwand near to the Ober See. The views throughout this walk are admirable, and it may well be recommended to moderate walkers who do not attempt the more laborious excursions mentioned below.

In ½ hr. from the Kessel the boat will carry the visitor to *St. Bartholomä*, a very ancient chapel built on a promontory formed by the Eisbach torrent, which has here borne vast masses of débris from a ravine of the Watzmann into the lake. Hence the ancient name, Bartholomäussee, which was supplanted by the modern name only in the last century, when a royal hunting lodge was built here near the chapel. As elsewhere in this district the sycamore grows to a great size, and is here the prevailing tree. When not occupied by the royal owner this is used as an inn, where the keeper supplies dinners to visitors, sometimes including chamois, venison, and saibling (*Salmo Alpinus*), the most esteemed fish of the Bavarian lakes eaten either fresh or dried and smoked. The walls show portraits of unusually large specimens of these fish, and of other animals that have been taken in the neighbourhood. The keeper can often point out chamois resting on some ledge or crag of the surrounding peaks, visible to the unpractised eye only through a telescope. The keeper does not supply beds to strangers, but the mountaineer can obtain dry hay and accommodation superior to that of a châlet. Many strangers visit the so-called *Eis Kapelle*, a name given by the natives to the hollow vault, formed in summer under a large mass of snow lying in a cleft, called the Eisthal, running deep into the mass of the Watzmann. This is annually renewed by the accumulations from spring avalanches. The path is rough, and 1 hr. is required to reach the spot from the landing-place. A fall of rock in 1861 has rendered the access more difficult.

A more interesting excursion is that to the *Ober See*. To reach it the traveller goes by boat to the SE. corner of the lake. A low ridge, covered by the rough pastures of the Sallet Alp, little more than ½ m. in width, separates the Königs See from the Ober See, a small sheet of water about 1½ m. in length, surrounded by steep rocks. The pedestrian may here take the path mentioned above, leading to the Gotzenalm, and cause the boat to meet him at the Kessel; or, if the day be not too far gone, he may follow the ridge northward from the Gotzenalm, and descend thence direct to Berchtesgaden.

Among other easy excursions from Berchtesgaden the traveller may go by the carriage road to Ramsau, and with an early start, may go as far as the Seissenberg Klamm, returning on the same day (see Rte. B).

Instead of following the road to the village of Ramsau, an excursion may be made to the *Wimbachthal*, which opens on the left of the road nearly 6 m. from Berchtesgaden. At the opening, the Wimbach torrent has cut a cleft through the limestone rock, similar in character to those that abound elsewhere in this district. After passing through this the traveller emerges into one of the wildest glens of the Bavarian Alps. To the l. rises the Watzmann, to the rt. the Steinberg (8,595'), without a break from the floor of the valley. There being no intermediate shelving plateaux to restrain the torrents and avalanches, enormous quantities of débris have been borne down on either side, and there seems no *à priori* improbability in the supposition of some German writers, that the glen was once a lake, similar in character and parallel in direction to the Königs See, but since filled up with débris. In that case it must have been much narrower than its neighbour, as the opposite mountains leave no wide space between their bases. The extent of bare limestone débris, the masses of snow remaining from the spring avalanches, and the absence of trees, save some patches of the dark creeping pine, give this glen a singularly wild and dreary character. The main stream, and the torrents descending from the mountains, are all concealed, making their way underground beneath the coarse gravel débris; no sound is heard save the occasional scream of some bird of prey, or the shrill whistle of the marmot. About 4 m. from the opening of the glen is one of the numerous shooting-boxes built by the late king, 3,084 ft. above the sea. Some 2 m. beyond this the glen, whose previous direction was SSW., turns to SSE., thus keeping parallel to the basin of the Königs See, which also bends to SE. in the short glen occupied by the Ober See; and in rather more than 1 hr. from the so-called Schlösschen, the traveller reaches the *Griesalp* (4,398'). Above this the path divides. One faintly marked track bears to the l, and by that way the mountaineer, with a local guide, may traverse a ridge projecting southward from the Watzmann, descend through the Eisthal to the Eis Kapelle, and so reach St. Bartholomä. The expedition may of course be made in the opposite direction, but the effect is far more pleasing when the traveller, after passing several hours amid stern and rugged rock-scenery, suddenly returns to the inhabited world amid the beautiful scenery that awaits him as he descends towards the lake. It would, however, be expedient to arrange beforehand for his conveyance homeward by boat from St. Bartholomä. The other rather easier, but longer, path goes by the *Trischbülalp* (5,749'), and Sigeretalp, to the Unterlahner alp (Rte. D), whence the traveller may descend to the lake by the Schranbach fall, taking care not to be detained for the night in some uninhabited spot on the shore. The mountaineer who would explore the Steinerne Meer (Rte. D) may ascend from the Unterlahneralp to the Funtensee. From the nature of the soil and the lowness of the temperature, the flora of the Wimbachthal includes many of the species of the subalpine and alpine zones. *Aquilegia Einseleana* is common throughout the valley.

There are few places in the Alps which offer to the mountaineer so many mountain summits, all accessible within a single day, and presenting such varied views, as those surrounding Berchtesgaden. The more interesting of these are here briefly noticed.

The *Untersberg*, lying about the centre of the triangle connecting Salzburg, Reichenhall, and Berchtesgaden, has been referred to in the preceding pages in connection with the two first-named places, but the ascent is more frequently made from Berchtesgaden. It is not very favourably situated for a panoramic view, being less isolated from the higher peaks to the S. than the Staufen or the Gaisberg. It offers, however, many attractions to the naturalist, and its upper level exhibits many of the characteristics of the limestone high plateaux, of which the best specimen in this district is the Steinerne Meer. Some additional interest is derived from the importance attached to the mountain in the mythical tales and fairy legends of this part of Germany, founded in great part on the numerous caverns and clefts, whence issue strange noises caused by falling water, and also from the remarkable aspect of the mountain, conspicuous throughout the neighbourhood of Salzburg and Reichenhall. The outline has often been likened to that of the Egyptian sphynx, with the head turned towards Berchtesgaden and the opposite end to Salzburg. The SE. face, towards the Albe, is extremely steep, almost a sheer precipice, while on the opposite side, facing the Saale, the slope is gentler and in great part clothed with forest.

The rounded summit seen from Salzburg is called Salzburger Hohe Thron (6,089'), while the highest, called Berchtesgadener Hohe Thron (6,467'), is at the SW. end of the ridge. Although nearly level, the central plateau of the mountain is far from being easily traversed. In some parts it is intersected by deep clefts, wherein snow lies till late in the summer, reminding the mountaineer of glacier crevasses, and the intervening space is sometimes reduced to a mere rough wall of rock, along which he must pick his way with caution. Wherever there is a little soil the plateau is covered with the tortuous trunks of the creeping pine, forming a barrier that is crossed with fatigue and difficulty even by the most active walker. Unless he be proof against thirst the traveller, intending to ascend this as well as most of the other neighbouring mountains, will be provided with drink, as he must be prepared for the possibility of finding no water on the mountain. The few springs known to the herdsmen are, to use the expressive German term, *Hungerquelle* (famine springs), often reduced to a mere driblet, descending drop by drop, and not seldom completely dry in fine weather.

The best way from Berchtesgaden is to follow the road towards Reichenhall for 1½ hr. At the hamlet of Krainwies a path turns off to the rt., ascends through forest, and in 1 hr. of constant rise reaches the Nienbachthörl, a depression in the ridge connecting the Untersberg with the Hallthurn Pass, whence the traveller already obtains a wide and interesting view in both directions. The way now turns to the rt. The ascent is at first through forest, then between rocks, where the sun beats with force, and in 1 hr. from the Thörl the Untersberger Alp is reached. Here those who wish to devote a long day to exploring the mountain may find night quarters. Near at hand is a rocky point commanding an extensive view. The châlets stand on a sort of promontory extending from the main mass of the mountain, and 2 hrs. are yet needed to reach the summit. After another ½ hr., chiefly through forest, the rough path reaches a sort of hollow in the mountain, called Bärenloch. Here a faintly marked track joins our path. It is the most direct way from Berchtesgaden, but is so rough and steep that it is rarely chosen. Here the view of the precipices of the Untersberg is extremely striking; the ascent becomes steeper, and the creeping pine takes the place of other coniferous trees. Continuing the ascent, with a view on the l. hand towards Reichenhall, the traveller before long attains the upper level of the mountain. The highest summit lies to the rt. and seems easy of attainment, but the numerous and deep clefts in the limestone form a labyrinth which severely tests the local knowledge of the guide. He cannot here, as on a glacier, extricate himself from a difficulty by a few well-directed blows of his axe, and it is often necessary to turn about and force a passage in a new direction; but the distance is not great, and before long the traveller attains the summit. The view of the ranges from the Dachstein to the Lofer Alps, and especially of the nearer mountains surrounding the Königs See, is extremely fine; but in the opposite direction the long ridge of the mountain itself conceals the city of Salzburg and a considerable portion of the plain. It appears as though it could be no more than a short walk to follow the ridge from the highest summit to the Salzburger Hohe Thron, whence one may look down directly on the city, and the course of the Salza and the Saale to the junction of those streams; but the difficulties of the way are such that from 4 to 5 hrs. of hard work are required to go from one point to another.

One of the curiosities of the mountain is the ice-cavern called *Kolowrats-Höhle*, discovered in 1846. The entrance is on the east side of a projecting hunch in the ridge, called the *Griereck*. This is scarcely accessible except from Glaneck (§ 44, Rte. A) a village 4 m. S. of Salzburg. From that place, where guides, ropes, and lights may be procured, it is reached in little more than 3 hrs.

To the botanist the flora of this mountain is very interesting, but owing to the broken nature of the surface, there is some uncertainty as to finding several of the rarer species. Among those characteristic of this region of the Alps may be noted *Saxifraga Burseriana*, *Pleurospermum Austriacum*, *Saussurea pygmæa*, *Soyeria montana* (this, as well as the last, only at the S. end of the mountain), *Willemetia apargioides*, *Campanula alpina*, *Tozzia alpina*, *Pedicularis incarnata*, and *Allium Victorialis*.

The *Hohe Göll* (8,266′) is the highest summit of the ridge dividing the basin of Berchtesgaden from the valley of the Salza. Its form at a distance, from whatever side it be seen, is that of a dome or cupola, though on a nearer approach it is found to be more sharply cut than had been supposed. Four ridges radiate in opposite directions from the peak, and between them are four hollow bays or amphitheatres, which give to this mountain a peculiar character. It has been unduly neglected by travellers, because of the supposed difficulty and even danger of the ascent, which are insisted on even in the recent edition of Schaubach's usually accurate work. Though a steep climb, there is nothing that will be considered a difficulty by a practised mountaineer. It is accessible from the E. by the Rossfeld, on the N. side of the Blüntauthal, or from Berchtesgaden by the Jagerschneid. Herr Hinterhuber, who made the ascent by the latter route, recommends Hinterbrandner of Berchtesgaden as a competent guide, and warns future travellers that very unreasonable demands were made at the Vogelalm, where he passed the night, a very unusual circumstance in the German Alps. These are the highest châlets, about 2½ hrs. from Berchtesgaden. The way thence lies by a rough slope called Jagerwiesel to the Jägerschneid ridge. The summit may be reached either by following the ridge of the *Hohe Brett* (7,690′), which is a secondary peak of the Hohe Göll, or else by a more direct but very rough course through the Todten Graben, one of the four hollows or recesses spoken of above. The highest point is marked by a wooden cross. A somewhat lower projecting peak is marked by an iron cross, beside which is a little box with a book, wherein the few who reach the summit inscribe their names.

A passage called the Rauchfang, formerly frequented by poachers, has been rendered quite impassable. Besides other rare plants, the botanist may find here *Valeriana Supina*.

The *Jenner* (6,162′) is a pyramidal summit projecting westward from the range connecting the Hohe Göll with the Hagengebirge. Unlike the Göll, it is very easy of access, yet commands a remarkably fine view of all the neighbouring mountains, besides which it looks down into most of the surrounding glens. The traveller may descend towards the Königs See by the Königsbach, joining the path of the Torrener Joch (Rte. F); or he may make a rather longer detour, with a guide, and descend to the Kessel, taking care to have a boat in readiness to carry him to his inn.

The *Schneibstein* (7,422′) lies in the main ridge overlooking the valley of the Salza, at the head of the Blüntauthal (Rte. F.), about 3 m. S. of the Hohe Göll. The view is in some respects similar to that from the latter peak, but less extensive. It is considered by Bavarian botanists to be one of the best habitats for rare plants. Besides most of the species found on the other peaks of this district, *Cherleria imbricata, Phaca frigida*, and *Sesleria tenella*, have been found here.

On the S. side of the Schneibstein extends a considerable mountain mass collectively called *Hagengebirge*, whose western headland is the *Kallersberg* (7,709'). Next to the Göll this commands the finest general view of any of the summits E. of the Königs See. The rocks contain numerous fossils. The most convenient way to approach it is from the Mitterhütte (about 5,200'), where the traveller may sleep after having, on the previous day, ascended the Jenner or the Schneibstein. Thence he ascends to the Kallersberger Alp, the highest châlets in this district, and in 2 hrs. more attains the summit. Instead of shattered rocks, absolutely bare, or covered only with lichens, and some little flower rooted here and there in a cleft, the peak presents a slope clothed with alpine vegetation, reminding the mountaineer of the ordinary aspect of the Swiss and Savoy Alps. In descending he may take a very interesting way from the Kallersberger Alp by the *Bärensunk*, a singularly wild hollow, half filled with fallen rocks, where the herdsmen obtain their supply of water only by melting the snow remaining from the spring avalanches. By the so-called Steinerne Stiege a track leads to the Landthal, a trough dividing the Hagengebirge from the lower summits overlooking the Königs See. Here the traveller may choose between the way by the Gotzenalm to the Kessel, or keeping more to the l., he may reach the Regenalm, and descend by the Kaunerwand to the head of the Königs See. Should the love of the wildest alpine scenery induce him to spend another night in some herdsman's hut, he should turn to the l. from the Landthal, and follow a track called Luchspfad, which will lead him up to the plateau of the Hagengebirge, intermediate in character and in height between that of the Untersberg and the Steinerne Meer. Numerous summits rise from 500' to 1,000' above the general level. The plateau is bounded on the N. by the Blüntauthal, on the E. by the defile of Lueg in the valley of the Salza, and to the S. by the Blühnbachthal. Even though the traveller should not intend to descend into the latter valley, he should not fail to reach the pass of the *Wildthor* (Rte. F.), and gain the grand view of the Uebergossene Alp, which there rises exactly face to face. On the way, going or returning. he should make a slight detour to attain the ridge of the Laubsattel, whence he looks down on the Ober See and the upper end of the Königs See.

The *Steinerne Meer*, which, with its alpine lakes and the impressive dreariness of its high plateau, may well detain the mountaineer for two or three days, is noticed in Rte. D. Those who do not intend to cross to Saalfelden may reach one or other of its higher summits—*Funtenseetauern* (8,388'), or *Hundstod* (8,532')—from St. Bartholomä, or from some sennhütte that would be reached on the first day from Berchtesgaden. The return to that place may be made on the following day by the Wimbachthal.

The *Watzmann* (8,988'), being the highest summit visible from Berchtesgaden, has naturally attracted the attention of many mountaineers. Deservedly so, for the view is very grand and extensive; but the ascent does not afford as much of incident and variety as that of many other peaks of lesser height. Though at first sight it appears very steep, the ascent is free from all difficulty, and reduces itself to a long, nearly uniform climb along an arête, not narrow enough to trouble any except persons predisposed to giddiness. To the practised mountaineer a guide is not necessary, so long as he keeps to the northern arête; but in this, and all other mountain excursions in this district, the tendency of the rocks to split into deep clefts should be borne in mind. The stranger is liable at every turn to be stopped by some impassable chasm where he anticipated no difficulty in his way. In order to reach the summit, it is customary to sleep at

ROUTE A.—ASCENT OF THE WATZMANN.

one of the châlets on the northern slope of the mountain. The usual, and the easiest, course is to commence the ascent from the road to Ramsau, a short way before the junction of the Wimbach with the Ramsauer Ache. After mounting for nearly 2 hours through forest, the path enters on alpine pastures, broken by projecting rocks and scattered blocks. It is usual to pass the night at the *Guglalp* (5,139′), which offers worse accommodation than is usually found in the Bavarian Alps. Above the *alp* the way becomes rather steeper, the broad rounded ridge becomes gradually narrower, and from time to time the traveller approaches the verge of the precipices facing the Wimbachthal. To the l. is seen a hollow filled with snow, called the Dürre Grube, visible from Berchtesgaden, and above this, in 2 hrs. from the Guglalp, a point in the ridge is attained whence for the first time the summit is visible. On the opposite side of the Dürre Grube another similar ridge meets that by which the traveller has hitherto ascended, and by that way the track from Königssee, mentioned below, joins the ordinary path. Henceforward the two routes are united. The ridge becomes more and more narrow as it rises towards the peak, but it is only where it approaches the summit that even nervous persons can find any difficulty. In from 3½ to 4 hrs. from the alp the traveller attains the Hohe Spitze, marked by a large stone man and a painted cross. The cairn covers most of the narrow level space, leaving little room for visitors. A book is deposited in a recess, wherein they write their names.

The Watzmann is a double peak, and the southern, or Hintere Spitze, appears to be higher by a few ft. This is sometimes called Schönfeld Spitze, but is not to be confounded with the highest point of the Steinerne Meer. The ridge connecting the two summits is so rent by deep clefts that it is a tedious and difficult operation to pass from one to the other. It appears that the S. peak has been reached but twice, and the passage involves some risk. Overlooking all the mountains within a radius of more than 30 m., excepting only the Uebergossene Alp, the view offers a very extensive and grand panorama, wherein the main chain of the Noric Alps chiefly attracts the view. The nearer views are scarcely less interesting.

In the writer's opinion, the way from the hamlet of Königssee, though rather steeper, is decidedly more agreeable than that by the Guglalp. An active mountaineer will have no difficulty in making the ascent from the inn at that place, but it is true that the peak is more often clear at an early hour; and unless he start from the inn considerably before daylight, he will do better to sleep on the mountain. The path from Königssee mounts steeply on one side of the gorge of the Klingerbach to the large and clean-looking châlet of Herrenrain (4,163′), thence bearing to the l. to the Kührainalp. This stands on a ridge running parallel to the main mass of the Watzmann, which culminates towards SSW. in a peak called the *Kleine Watzmann*. Below the Kührainalp this ridge is separated from the slope leading up to the Guglalp by a short glen, called the Schapbachthal, through which a torrent runs down to the Ramsauer Ache. Above the alp, a wild, stony, trough-shaped hollow, called the Watzmannscharte, runs up between the ridge of the Kleine Watzmann and its loftier namesake. Towards the summit of this hollow is a small glacier. The lower part of the Watzmannscharte, opposite the Kührainalp, is nearly level, and in order to reach the main peak, the traveller crosses it diagonally, and ascends by the Mitteralp to the *Fatzalp* (5,506′), where he may pass the night as well, or as ill, as at the Guglalp. A stiff climb then takes him up the ridge at the l. side of the Dürre Grube, above which he joins the ordinary route to the summit.

If the traveller ascending from Königssee should find clouds gathering round the summit of the mountain, he will perhaps do well to alter his course, as did the writer under such circumstances, and turn his steps to the Watzmannscharte. As he ascends, the rocks of the Great and Little Watzmann rise more and more steeply on either hand, and the scene, when he reaches the little glacier at the top, is one of the wildest that can be imagined; the effect being heightened, if, through some break in the clouds, the green meadows, and luxuriant foliage, and neat houses of the vale of Berchtesgaden are seen in strange contrast with the savage scenery that surrounds him. Most of the rarer plants mentioned below are to be gathered here. The botanist will reap an ample harvest on this mountain. To mention but a few of the more interesting species, he will find in the lower alpine zone (5-6,000') *Heracleum austriacum, Senecio abrotanifolius,* and *Rhododendron chamæcistus;* and in the upper zone, *Papaver alpinum* (on the highest peak), *Draba Sauteri, Cherleria imbricata, Potentilla clusiana, Saxifraga stenopetala, Valeriana supina, Leontodon Taraxaci, Crepis hyoseridifolia, Campanula alpina, Primula minima, Kobresia caricina,* and *Sesleria tenella.*

With this fine peak we close the list of the excursions from Berchtesgaden, as the Hochkalter and the Kammerlinghorn are more conveniently visited from Hirschbühel. (See next Rte.)

The botanist at Berchtesgaden should apply to Herr Apotheker Birnguber for permission to inspect a manuscript catalogue of rare plants found in the neighbourhood by the late Dr. Einsele.

ROUTE B.

SALZBURG TO ZELL-AM-SEE, BY BERCHTESGADEN.

	Post Stunden	Eng. miles
Berchtesgaden	6	14
Ramsau	2¾	6½
Ober Weisbach	6	14
Saalfelden	4½	10½
Zell-am-See	4	9½
	23¼	54½

The road here described, of which one portion is fit only for light country carriages, is the most interesting way for a traveller going by road from Salzburg to the Upper Pinzgau, being superior in scenery, and nearly as short as the post-road by Lofer. It may also be recommended to those going from Salzburg to Gastein, who wish to visit Berchtesgaden on the way. The scenery on the way from the latter place to Zell-am-See is of a very high order, and includes one of the most singular ravines in the Alps. The water of the *Albe*, also called Ache, and Alm, mentioned in the last Rte. as the stream which drains the basin of Berchtesgaden into the Salza, is in part diverted into a canal, which is carried across the plain NE. of the Untersberg to the city of Salzburg. The road to Berchtesgaden keeps for several miles near to this canal, leaving to the rt. the village of Glaneck, at the base of the Untersberg. After passing Gredig, a projecting rock that looks like a huge block fallen from the Untersberg, half closes the entrance to the valley of the Albe. On its south side that stream turns eastward towards the Salza, and a branch road runs along its l. bank to join the high road from Salzburg to Hallein (Rte. F.) Near the point where the canal is derived from the Albe are some considerable marble quarries and saw mills, established by King Louis of Bavaria. The valley soon contracts to a defile between the steep SE. face of the Untersberg and a ridge projecting northward from the Hohe Göll. The name *Hangende Stein* is sometimes given to the defile in general, sometimes to the point where the frontier between Austria and Bavaria was fixed in 1815, when Berchtesgaden was preserved to Bavaria, while the secularised territory of Salzburg was assigned to Austria. This was the ancient limit of the little territory of Berchtesgaden, which long preserved a separate existence under an ecclesiastical ruler, who bore the title of prior, in spite of many attempts on the part of successive prince-archbishops of Salzburg to gain possession of it by force or by diplomacy. There is still seen the inscription, 'Pax intrantibus et inhabitantibus,' set up at all the frontiers of his little territory by a certain Prior Rainer in 1514. Soon after passing the Bavarian customs stations, the road reaches

Schellenberg (1,506'), a large village most picturesquely situated. The road now crosses the stream, keeping that side till near Berchtesgaden. On issuing from the defile, the traveller gains at length a view of the mountains surrounding Berchtesgaden, among which the Watzmann towers supreme. To the l. is seen a country road practicable for light carriages, which goes to Hallein over a deep depression in the ridge, on the E. side of the Albe. Passing opposite to the Salzberg, the traveller soon after reaches

Berchtesgaden. As mentioned in last Rte., the road to Saalfelden follows the SW. branch of the Albe, henceforward called Ramsauer Ache. For more than 4 m. the pipes conveying saturated brine from the Salzberg to Reichenhall run alongside the road. The reason for taking a direction so wide of the shortest line leading to that place is apparent when, on reaching the Ilsangmühle, the traveller sees the machinery by which the brine is conveyed over the pass of Schwarzbachwacht (2,907'). At this point there existed a vast amount of mechanical power, embodied in the fall of a mountain torrent from a height of 384 ft., uselessly expended until the local engineers thought of setting it to the task of raising the brine over the ridge dividing the Albe from the Saale. The lowest pass over that ridge is that of Hallthurn, only 2,225 ft. above the sea (Rte. A.); but it was found expedient to lengthen the route by many miles, and to traverse a pass nearly 700 ft. higher, in order to get the benefit of an agent costing next to nothing, yet able to raise the brine to a height of 1,263 ft. above the reservoir wherein it is received at the Ilsangmühle. After passing the opening of the Wimbachthal, noticed in the last Rte., the road reaches

Ramsau (2,163'), a small village with a good country inn, a chosen resort of Munich landscape painters. The village is in the immediate neighbourhood of much pleasing and even grand scenery, but the site is not equal to that of Berchtesgaden. The forms of the Watzmann and other peaks as seen from this side are far less bold, and there is not so varied and rich a foreground. Those who do not traverse the whole road from hence to Reichenhall, described in the next Rte., should go at least as far as the Taubensee, and may on the same day make a pleasant detour by the Reiter Alp, either to the Reitalphorn (5,747'), overlooking the valley of the Saale from Lofer to Reichenhall, or to one or other of the peaks of the Reiter Steinberg mentioned below.

About ½ m. beyond the village, the road leading to Reichenhall turns off to the rt., while that to Saalfelden preserves the westerly direction which it has held nearly all the way from Berchtesgaden. After traversing some outcropping beds of triassic Bunter sandstone the limestone reappears; and the road, after twice crossing the torrent, reaches the *Hintersee* (2,603'), one of the wildest and most picturesque of the many mountain lakes of the Bavarian alps. It occupies a deep depression between the range of the *Steinberg*, culminating in the *Hochkalter* (8,595'), and a still more rough and boldly sculptured mass, sometimes collectively called the Reiter Steinberg, and including several summits, the highest of which are the *Mühlsturzhorn* (7,470'), the *Stadlhorn* (7,449'), and the *Spitzhörndl* (7,229'). Among the few houses by the lake shore is a very fair country inn and a hunting lodge of King Max. It would appear as if the SW. end of the lake had been filled up with débris fallen from the surrounding mountains; and no stream is seen to flow through the barren stony plain, over which the road is carried, till it reaches a rocky barrier, which has to be surmounted in order to gain the head of the valley. Gaining a view of the *Hochcisspitz* (8,259'), at the SW. end of the range of the Steinberg, the road finally ascends to the *Hirschbühel Pass* (3,896'), here marking the frontier between Austria and Bavaria. A small but good country inn stands beside the customs station.

This is well situated for the mountaineer who would explore the Steinberg range. The ordinary excursion is the ascent of the *Kammerlinghorn* (8,146'), easily reached in 3½ hrs., and commanding a very fine panoramic view. The botanist may find *Orchis Spitzelii* and *Draba Spitzelii* during the ascent.

The pedestrian wishing to take the shortest way to Lofer may follow a path to the rt., which crosses the *Kleine Hirschbühel* (4,243'), and decends by Wildenthal to the high road above St. Martin. The road to Saalfelden—fit only for very light carriages—descends at first by the rt., then at the l. side of the *Weissbach* torrent. About 2 m. from the pass a finger-post by the road side points the way by which the traveller may descend into one of the most extraordinary of those deep clefts that have been cut by mountain torrents through the limestone strata in this part of the Alps. This is known as the *Scissenberg Klamm*. The pathway is for the most part a wooden stage, erected for the purpose of clearing the masses of felled timber that are floated down the torrent for conveyance by the Salza to Reichenhall. In one part the direct light of day never penetrates into the chasm, where the torrent rages and frets against the uneven surface of the rock; and the faint light reflected from the rocks is farther softened by the bushes rooted in the clefts above that partly roof over the opening. Excepting the Schwarzenberg Klamm, near Unken (§ 44, Rte. A.), this is the most remarkable of the ravines of this district, and may almost bear comparison with the famous gorge of Pfäfers, in Switzerland. The road enters the valley of the Saale at the scattered village of *Ober Weissbach*. The good country inn (zur Frohnwies), 2,173 ft. above the sea, is at the upper end, some way from the church. The charge for a light one-horse carriage from Frohnwies to Berchtesgaden is 6 fl.

From Frohnwies upwards, for a distance of nearly 6 m., the valley of the Saale forms a remarkable defile, the passage of which in bad weather is not altogether free from risk. Through the *Hohlweg*, as it is called, the floor of the valley is nearly level and of moderate width, but it is enclosed on either side by precipices which at many points are absolutely vertical. In the spring, and at other seasons after wet weather, masses of rock are often detached, and large fragments not unfrequently reach the road. The slight flavour of danger, and the sternness of the scenery, add interest to the passage. At one point, where the Diesbach torrent descends in a waterfall from the Steinerne Meer to join the Saale, stands a solitary mill, the only house in the defile. Even the lover of fine scenery feels relief when, at a point where the valley bends from SSE. to due S., the deep trench widens out into a comparatively wide green basin backed by the snowy peak of the Vischbachhorn, beyond which the summit of the Glockner is also visible. In the midst of this basin, 2,476 ft. above the sea, stands *Saalfelden* (Inns: Auerwirth; Timmerlwirth). The village stands on the Urschlauer Ache, nearly 1 m. E. of its confluence with the Saale. The position is extremely fine. The valley of the Saale is here intersected by a line of valley (mentioned in the introduction), which to the E. forms the *Urschlauthal* or Urslauthal (Rte. G.), and to the W. the Leogangthal, noticed in § 44, Rte. F. To the NE. the stern range of the Steinerne Meer (Rte. D.) rises in menacing proximity to the village, while from many points in the valley the snowy points of the Glockner and its attendant peaks form an admirable background. Numerous castles, each with its store of historical or legendary recollections, crown the eminences on either side of the valley. The Lichtenberg is that which commands the finest view. Above Saalfelden the broad valley of the Saale extends for several miles nearly at a dead level. The road, after crossing the stream, runs due S. on the W. side of the valley, the floor of which is broken into rough undulating ground, covered by trees, and the stream seems at no distant time to have changed its channel. When at *Saalhof*, a castle about 7 m. above Saalfelden, a stream flowing from the W. is crossed by a wooden bridge, while the road runs straight on to the lake of Zell, no stranger could suspect that in the short level space thus traversed he had crossed a watershed between one river-basin and another. Such, however, is the fact. The stream left on the rt. hand is the Glemmbach, the main source of the Saale (§ 44, Rte. G), while the lake sends its waters southward to the Salza. Careful examination is needed to ascertain whether the lake did not at one time extend from the Salza valley to Saalfelden. It would then have received at once the head waters of the Saale and the Salza, and the outflow must have been carried, either at one and the same time, or alternately, through the valleys now traversed by both those rivers.

Leaving to the l. the castle of *Prielau*, standing on swampy ground, at the N. end of the lake, and now used as an inn, the road runs along the W. shore to

Zell-am-See (Inns: Bräu, clean and fairly good; Lebzeltner, reasonable), a village little resorted to by strangers, yet enjoying a position not surpassed by many others in the Alps. The lake from which it takes its name—the *Zeller See*—lies about half-way between the highest portion of the main range of the Noric Alps and the extremely bold and rugged mass of the Steinerne Meer and Uebergossene Alp. In its usually placid surface both the snowy range to SSW., and the grey limestone masses to NE., are mirrored. The lake is said, on the authority of L. von Buch, to be 620 ft. in depth, and its height above the sea is about 2,450 ft.; that of the village being 2,469', or nearly exactly that of Saalfelden. The traveller should not fail to take a boat excursion on the lake, which is more than 3 m. long, and nearly 2 m. wide. The views in both directions are finer than from any point on the shore. In the lake are found two very rare water-plants, *Nymphæa biradiata* and *Ægrophila Santeri*.

Of mountain ascents most to be recommended is that of the *Hundstein* (6,946'), reached in 4 hrs. from the village. The view is of a high order, and it is one of the best points for studying the great SW. face of the Uebergossene Alp, which on this side is called Wetterwand. *Gentiana prostrata* has been found on the mountain. Nearer at hand is the *Schmittener Höhe* (6,441'), lying W. of Zell-am-See, and easily reached in 3 hrs. Still nearer, but not commanding so good a view of the lake, is the *Hönigkogl* (6,085'). There is one serious drawback on the advantages of Zell-am See as head-quarters for alpine tourists, in the neighbourhood of the extensive marshes, which spread widely from the S. end of the lake into the valley of the Salza. The village is, indeed, supposed to be free from the malaria caused by them, but they undoubtedly tend to disfigure a site which in other respects offers many attractions.

In spite of its small size, Zell-am-See is a local centre of some little importance. Country diligences ply twice a day in 4 hrs. to Mittersill, another runs to Saalfelden and Lofer, and another to Taxenbach and Lend. The church of St. Hippolytus contains some curious work in carved stone, and deserves a visit.

Route C.

SALZBURG TO ZELL-AM-SEE, BY REICHENHALL.

It being supposed that most travellers going from Salzburg to Zell-am-See desire to pass by Berchtesgaden, precedence was given to that way in the last Rte. The easiest and shortest way is, however, by Reichenhall, along the good road which nearly all the way keeps to the valley of the Saale, and which, though offering less variety, yet affords the traveller an opportunity of seeing some very fine scenery. A third way, best suited for the pedestrian, combines a visit to Reichenhall with some of the most striking scenery described in the last Rte.

1. *By the valley of the Saale.*

	Post Stunden	Eng. miles
Reichenhall	4	9¼
Unken	5	11⅔
Lofer	2¼	5¼
Weissbach	2½	6
Zell-am-See	8½	20
	22¼	52¼

The road between Salzburg and Lofer is described in § 44, Rte. A.

At Lofer the traveller may find a post omnibus that plies daily to Zell-am-See, but unless he secures a front place he will see next to nothing of the scenery.

Above Lofer the road runs for some way over marshy ground, and according to local tradition, passes at a place called *Gumping*, the site of a town now buried in the earth. The occasional discovery of objects of antiquity has given some colour to the popular belief. A little farther is the village of *St. Martin*, with a fine gothic church, opposite the opening of the *Kirchenthal*, a short glen descending from the Ochsenhörner. Its torrent, after reaching the valley, runs parallel to the Saale, and joins that stream only close to Lofer. The pedestrian who would gain a fine view may turn aside, and mount by a paved track to the little village of *Kirchenthal* (2,898'). The church, a handsome building in the Italian style, is frequented by pilgrims.

A short distance beyond the village of St. Martin the valley is narrowed between the E. base of the Loferer Steinberg, on the l. bank, and the Reiter Steinberg, also called Reiteralmgebirge, that rises above the rt. bank of the Saale. The defile, called *Pass Luftenstein*, was formerly fortified, like most of the similar defiles in this district. Here the path from the Kleine Hirschbühel, mentioned in the last Rte., descends into the valley. On a rock above the pass is the ruined castle of Lambrecht, near to which is a cavern, called Lambrechtsofenloch, to which is attached a legend that has made it the frequent resort of the natives of the valley, bent on the fruitless search for buried treasure. Less than a mile farther is the hamlet of Unter Weissbach, followed by that of Ober Weissbach, where the road from Berchtesgaden by the Hirschbühel Pass joins our route. The remainder of the way to Zell-am-See is described in Rte. B.

2. *By the Taubensee.* It was mentioned in Rte. B that instead of carrying the brine fom the salt mine at Berchtesgaden to Reichenhall by the shortest road, and over the lowest pass between those places, the pipes are led along the valley towards Ramsau, and the brine then raised by water-power to the pass of Schwarzbachwacht, whence it descends to Reichenhall. There is a rough road from Reichenhall to Ramsau, for the most part following the line of the brine-

conduits, which may be taken on the way to Saalfelden. It is practicable for light carriages, but is better suited for the pedestrian. The distance from Reichenhall to Ramsau is about 14 Eng. miles, and avoiding the latter village, the distance to Weissbach is about 27 m., or 4 m. more than by the high-road. This road leaves Reichenhall by the rt. bank of the Saale and follows the stream for more than 5 m. On the l. hand is seen the steep northern face of the *Lattenberg*, which is the collective name for the mass of mountain, rather a high plateau than a range, that divides the Saale from the basin of Berchtesgaden. Several summits, the highest of which is the *Dreysesselkopf* (5,530'), rise above the general level of the plateau, which is covered with rich pasture, being, unlike the other mountains of this district, well supplied with water. About 3 m. from Reichenhall the road crosses the Röthelbach, a stream descending from a tract of peat moss near the summit of the plateau. By that way goes a path frequented by the herdsmen. Nearly 2 hrs. from Reichenhall the road reaches the opening of a more considerable stream— the *Schwarzbach*—issuing from a glen that divides the Lattenberg from the much higher and more rugged mass, collectively known as the *Reiter Steinberg*. Through this glen mounts the road to Ramsau, passing at its opening the little village of

Jettenberg (1,626'), lying close to the foot of the very bold summit of the Alphorn (5,716'), which projects northward as a promontory from the mass of the Steinberg. From the village the road ascends for 2 hrs., often passing alongside of the brine pipes, to the *Schwarzbachwacht* (2,907'), being the summit of the pass, and the site of the reservoir where the brine is received after being forced up from the Ilsangmühle (Rte. B). The descent on the S. side is much shorter than the ascent from Jettenberg, and the scenery very interesting, commanding very fine views of the Watzmann, and other surrounding mountains. To the rt. rise, behind the outer slopes of the Reiteralp, some of the higher summits of that mass.

[If accompanied by a guide, the traveller may find a way across the range to the valley of the Saale, descending to the hamlet of Reit, on the rt. bank of that stream between Unken and Lofer.] Leaving the brine pipes to the l., the road to Ramsau descends to the *Taubensee*, a highly picturesque little mountain lake with a few rocky islets, and in ½ hr. more joins the road between Ramsau and Hirschbühel about ½ m. from the former village. This is avoided by the traveller bound for Saalfelden and Zell-am-See, who turns to the rt., and follows the road described in Rte. B.

The summits of the Reiter Steinberg, and especially the E. point, locally called Eisberg, are unusually rich in rare plants.

Route D.

BERCHTESGADEN TO SAALFELDEN, BY THE STEINERNE MEER.

The roads described in the two preceding Rtes. all make a considerable detour round the mass of high mountains enclosing the Königs See; but a glance at the map shows that the direct way from Salzburg and Berchtesgaden to the upper valley of the Saale is over the rugged mass of mountain called *Steinerne Meer*, lying S. of that lake. Except by practised mountaineers the distance from Berchtesgaden to Saalfelden will be found a very long day's walk, and it is advisable to sleep at the inn at Königssee, and start by boat at the earliest dawn, unless the traveller should prefer to sleep at a châlet. The latter alternative is recommended to the naturalist, who will thus gain time to enable him to examine the singular region over which his track is carried. The usual way to the Steinerne Meer is to land beside the *Schranbachfall*, a waterfall on the SW. shore of the Königs See. The spray falls so widely into the lake as to wet persons approaching the spot in a boat. A steep ascent of ½ hr., partly through forest, leads to the Schranbachalp (2,859′). Beyond this the copious stream of the Schranbach, which lower down forms the waterfall, disappears from view. The track crosses the bed of a little lake that seems to have been drained through some subterranean channel. Once, or oftener, during the ascent the torrent will be seen to emerge to daylight, and then again be either lost amidst the limestone gravel or drawn off through a fissure. From the *Unterlahneralp* a path turns off to the Sigeretalp and Trischübl (5,749′), whence the traveller may descend through the Wimbachthal to Ramsau, or return to Berchtesgaden. The Unterlahneralp (3,375′) appears to be completely enclosed towards the S. by an amphitheatre of limestone rocks, but there is a passage in that direction through a cleft called the *Saugasse*, in which lies a steep pile of fine débris mingled with large blocks. An hour's steep ascent is required to reach the summit, passing about half-way a spring of very cold water, possibly the last that the traveller will see for many hours, unless his guide should point out another to the l. of the track near to the *Oberlahneralp* (4,606′). This stands on a shelf of the mountain whereon he lands after passing through the Saugasse. Here the rocks begin to exhibit the crevassed character, but the clefts show a vigorous growth of alpine plants. Several circular hollows in the rock are passed on either hand, and here and there planks are laid over the wider crevasses. Still ascending, and passing through a sort of gap, called the Ofenloch, the traveller reaches in 3½ or 4 hrs. from the shore of the Königs See, the *Funtensee*, a somewhat dreary tarn,

5,248 ft. above the sea. Keeping to the l. he reaches a point where the rocks descend almost vertically to the water's edge, and a rumbling and gurgling sound is heard under the surface. The tendency of floating bodies thrown into the water towards this spot shows that the little lake has here a subterranean outlet. Rugged limestone summits rise on every side, the highest of which—the *Funtenseetauern* (8,388′)—lies to SE. The châlets at the farther end of the lake are tolerably clean, and the traveller may there find tolerable night-quarters. If his guide be not thoroughly acquainted with the pass, it will be prudent to take some one from the Alp to show the way over the trackless plateau of the Steinerne Meer. From the châlets, or from those of Im Feld (mentioned below), several of the higher summits may be reached, or the naturalist may devote a long day to exploring the stony desert. The Funtenseetauern is reached in about 3 hrs., passing by the Feldalph. The highest summit—the *Schönfeldspitze* (8,696′)—has been rarely attained.

The way to Saalfelden ascends between the *Vichkogl* (7,080′) and the *Hirschkopf* (6,360′) to the *Schönbühlalp* (6,114′), a small green tract—the highest pasture—almost surrounded by the bare rocks, which from a distance simulate in a strange way the aspect and form of a glacier. Here commences the passage of the tract to which the term Steinerne Meer (sea of stone) properly belongs, though the name is commonly given to the entire mountain mass lying S. of the Königs See. The high plateau which the traveller now traverses is not by any means a solitary example of the kind, as numerous other instances may be found in the Styrian and Carnic Alps, not to mention that of the Untersberg, described in Rte. A; but this is probably the most remarkable for its extent, and for the absolute nudity of the surface. Its length, from the *Seehorn* (7,416′) to NW., to the *Brandhorn* (8,035′) towards SE., is above 5 m., and its breadth in most places is not less than 2 m. In some seasons a trickling spring of water may be found at one or two points of the surface; but as a general rule it is absolutely devoid of moisture, and the keen eye of the botanist can scarcely discern the few rare specimens of alpine vegetation that nestle in the chinks of the rock. To the ordinary observer it is absolutely bare, silent as the grave; perhaps as absolute a desert as is to be found anywhere on the earth's surface, well deserving the name 'Todtes Gebirge,' given in Styria to such wastes. The rifted aspect of the rock, and the rounded hollows that recur at intervals, involuntarily recall the appearance of the moon's surface as seen through a powerful telescope. In $2\frac{1}{2}$ hrs. from the Funtensee the traveller reaches the *Weissbach Scharte* (7,462'). The plateau, which has hitherto sloped upwards to the S., here comes abruptly to an end, and the traveller has before him a view of unexpected grandeur. Standing on the verge of what appears a sheer precipice, he looks down upon Saalfelden, and the Zeller See, and the Dienten mountains, here dwarfed to mere hills; while behind them is spread out from W. to E. the great snowy chain of the Hohe Tauern from the Dreiherrnspitz to the Ankogl. Except the main range of the Pennine Alps, there is no continuous ridge in the Alps so unbroken by deep depressions. Facing round to the N. the view of the stony desert just traversed is scarcely less remarkable. The descent is extremely steep and rugged, chiefly by slopes of débris lying at the highest angle of repose, till in 1 hr. the *Weissbachalm* (5,681') is reached. Here spring water, and usually milk, may be obtained. A better path is now found. At the Maria Alm, about $1\frac{1}{2}$ hr. lower down, is a little inn where beer, bread, and eggs may be had; but 1 hr. more of easy walking carries the traveller to Saalfelden (Rte. B). Ten hours, exclusive of halts, should be allowed for the walk from the shore of the Königs See to that place.

A decidedly more interesting and rather shorter way than that above described is by the Grünsee. Instead of landing at the Schranbach waterfall the boat should be taken to the Sallet Alp, where visitors to the lake usually go ashore to visit the Obersee; thence ascending to the Grünsee. The following extracts are from notes kindly supplied by Mr. F. F. Tuckett:—'We quitted the boat at 3.30 p.m., and taking it leisurely but steadily, with only occasional halts to admire the exquisite views, we followed for the next 2 or 3 hrs. one of the most wonderful mountain paths I ever saw. It was a mixture of the Gemmi and the Dala Leitern, enriched by a beautiful growth of trees and ferns, and by glorious views of the lake; so close, that although some 2,000 ft. below us, we could apparently have chucked a stone into it.' The steep face of the mountain up which the track is carried is called the *Sagereckwand*. After losing $\frac{1}{2}$ hr. in a fruitless attempt to make a short cut, Mr. Tuckett's party reached the Sagereckalp at 7 p.m., where the first good water seen since leaving the lake was found. This alp, which commands a noble view, is not occupied during the height of summer. Another $\frac{1}{2}$ hr. ascent over curiously broken ground brings the traveller to the beautiful *Grünsee* (4,504'). This lies in a hollow open to the N., but enclosed on the E. by the *Hohenscheibe* (7,950'), on the S. by the *Grünseetauern*, and on the W. by the *Simmetsberg*. Those who do not desire to reach the upper level of the Steinerne Meer may return to the Königs See by a path which winds round the Simmetsberg to the Oberlahneralp, and thence descends to the Schranbach fall. A short way above the Grünsee are the sennhütten of the Grünseealp, where fresh milk, and in case of need, rough quarters for the night may be obtained. 'From hence the path or track, if such it may be called, leads up the rugged wall of the Grünseetauern to a lonely group of châlets, called by the senner Im Fadel, but designated Im Feld on the Bavarian ordnance map. Here we received every attention and hospitality, and passed a most merry evening. Plenty of clean hay furnished us with excellent beds, and we got a capital night's rest.' The châlets of Im Feld, generally known here as the *Feldalp* (6,048'), are reached in about 1¼ hr. from the Grünsee, and in from 3½ to 4 hrs. from the Sallet Alp, or about the same time as is required to ascend to the Funtensee from the Schranbach fall, though they are fully 700 ft. higher. They lie in a depression of the ridge that separates the Funtensee from the much deeper hollow occupied by the Grünsee. The Feldalp commands a beautiful view, in which a portion of the Königs See is seen in marvellous contrast with the rugged and barren limestone masses near at hand. Near this point the attention of the traveller will be attracted by the strange forms into which the rock has been weathered. Instead of circular pits, which are seen elsewhere, we here have rounded conical eminences, somewhat resembling a beehive in form, split by diverging clefts and crevices. At some points these eminences are so near together that it is scarcely possible to pass between them. The flora of this and the upper plateau of the Steinerne Meer includes most of the rare species mentioned in Rte. A as characteristic of the mountains surrounding the Königs See. The Siberian pine, Zirbe, or arolla (*Pinus Cembra*) is rather common, especially on the heights surrounding the Grünsee. From the Feldalp a track leads down to the Funtensee, but the traveller bound for Saalfelden mounts directly to the plateau of the Steinerne Meer, and in 2 hrs. steady walking will reach the ridge of the Weissbach Scharte.

There are, at least, two other points whence it is possible to descend from the plateau towards the valley of the Saale, but they are both somewhat difficult, and fit only for practised mountaineers. The most easterly of these, lying nearly due S. of the Funtensee, is the *Buchauer Scharte* (7,490'). A very steep descent leads thence by the Kreutzhofalm (5,148') to the village of *Alm* in the Urschlauthal (Rte. G). The other passage lies considerably NW. of the

Weissbach Scharte. The way lies from the Funtensee towards the S. base of the *Hundstod* (8,532'); the summit connecting the mass of the Steinerne Meer with the Watzmann. Crossing the ridge of the *Diesbach Scharte* (6,679'), the traveller descends to the Diesbachalm, and then, keeping along the *Diesbach* torrent, falls into the Hohlweg (Rte. B) between Frohnwies and Saalfelden. He may also take a path leading NW. from the Diesbachalm to Hirschbühel.

In case the traveller, after exploring some part of the plateau, should wish to descend to Ramsau or to Berchtesgaden by the Wimbachthal, he may traverse a comparatively deep depression in the ridge connecting the Hundstod with the *Gjaidköpfen* (7,626'), and so reach the Trischüblalp, whence an easy path leads down to the Wimbachthal (Rte. A).

ROUTE E.

SALZBURG TO LEND, BY THE VALLEY OF THE SALZA.

	Austrian miles	Eng. miles
Hallein	2	9½
Golling	2	9¼
Werfen	2¼	10¼
St. Johann	2¼	10½
Lend	2	9¼
	10¼	49½

The above are believed to be the true distances, but in posting the stages from Golling to Werfen, and thence to St. Johann, are, or formerly were, charged 3 miles each. The valley of the Salza, from Salzburg to St. Johann, is one of those considerable breaches in the general disposition of the Alpine chain that has suggested to geologists a belief in the action of mechanical forces acting on a large scale upon the earth's crust. In the present instance the conformation of the mountain masses, through which runs this deep cleft, is so irregular, that it is difficult to argue securely as to their original direction before they were broken up by the working of dynamical and meteorological agencies; but when we fix our attention on the great line of valley running parallel to the crystalline axis of the Noric Alps, and see that the stream of the Salza escapes from this main valley by a course at right angles to its previous direction, through a defile between mountain masses that must at one time have been continuous, it is difficult to avoid the belief that the opening was originally effected by mechanical force, however it may have been afterwards enlarged by erosive action.

Like other similar transverse valleys, this has been turned to account for the purpose of running a high-road through the outer ranges of the Northern Alps into and across the central chain. From St. Johann the road described below turns westward, and is carried through the Pinzgau quite to the head of the valley of the Salza. Another road in the opposite direction leads to Radstadt, and thence either eastward into Styria, or southward into Carinthia; but the larger part of the traffic, at least in summer, arises from the concourse of visitors proceeding to the baths of Gastein. In the present Rte. the road to that place is described as far as Lend, but the Gastein valley is included in § 52.

In summer a diligence plies daily from Salzburg to Bad Gastein, a distance of 15 Austrian (70½ English) miles, in 13½ hrs. The fares are rather high, and for two persons travelling together it costs no more to hire a small carriage. The usual charge to Gastein, with one horse, taking 1½ day for the journey, is 12 fl. and 1 fl. trinkgeld. The scenery is in some parts fine, and always inter-esting, but not equal to that of the road by Hirschbühel and Saalfelden, described in Rte. B.

The post road from Salzburg keeps to the l. bank of the Salza, but there is an equally good road by the opposite bank. The former passes by the Castle of Hellbrunn (§ 44, Rte. A), and the modern castle of Anif, belonging to Count Arco. To the same proprietor belongs a very large brewery, formerly appertaining to one of the royal family of Bavaria, passed on approaching Hallein. Near that town many large barges, used exclusively for the conveyance of salt, are usually seen on the river. Clouds of smoke and steam from the salt-works announce to the traveller his approach to

Hallein (Inns: Post or Adler; Sonne; travellers are also received at the Baths, or Soolbade-Anstalt), an ancient town, 1,473 ft. above the sea, famous since the time of the Romans for the salt mine, which is opened in the hill close to the town, but extends a considerable distance underground. It appears that the deposit containing the salt, either in the crystallized state or diffused through the mass, is continuous with that forming the Salzberg near Berchtesgaden. The ancient name of the mine is *Tuval*, but it is more commonly known as the *Dürnberg*, that being the name of the hill wherein the galleries are bored. These are at five different levels, and are connected by shafts descending at a steep incline down which the miners and visitors slide from one level to that below it. Having obtained through the innkeeper a ticket of admission, the visitor ascends on foot or by carriage in ¾ hr. to the village of Dürnberg (2,531'), with a handsome church built of marble and commanding a fine view. The entrance to the mine is by a long gallery or tunnel, and at the end of it the visitor, first equipped in a suitable dress, descends the slide or shoot at an angle varying from 38° to 46°, which leads him to the next lower level. The mode of extraction is very simple. Fresh water is introduced into large reservoirs, and gradually converted into brine, by dissolving the salt diffused through the matrix. When saturated, the brine is conveyed to boiling-houses where, at the cost of a vast consumption of fuel, it is converted into salt. When the visitor has reached the lowest level of the mine, he is conveyed back to daylight close to the town of Hallein, by riding astride on a seat conveyed very rapidly on an underground railway, through a tunnel more than ¾ mile in length. The annual produce of salt is about 25,000 tons. The strangeness and novelty of the expedition to the mine attracts many visitors, including not a few ladies.

For 10 or 12 m. beyond Hallein the valley is open, and the level floor is from 1½ to 2 m. in width. The post-road is carried along the E. side of the Salza, but at some distance from its banks. Rather more than 6 m. from Hallein is *Kuchl*, with a decent inn (beim Auer) commanding an excellent view of the Hohe Göll (Rte. A.), which rises due W. The traveller who intends visiting the Schwarzbach waterfall should here leave his vehicle, cross the Salza, and follow an easily-found track leading him to the fall, and saving a detour of 3 m. The church of St. Nicholas is a guide to the stranger seeking the fall.

Golling (Inn : Post, probably the best on this road) is a village overlooked by an old castle, frequented by travellers on account of the interesting scenery in the neighbourhood. The chief attraction is the waterfall of the Schwarzbach, better known as the *Gollingerfall.* Though the volume of water is not very considerable, this must be reckoned amongst the most picturesque of alpine waterfalls. It well deserves a visit, for which 1½ hr. suffices, or 1 hr. if the traveller hire a one-horse car, costing only 1 florin. The pedestrian, leaving the high road at Kuchl, as mentioned above, will lengthen his walk by little more than ½ hr. The *Schwarzbach* torrent is with great probability believed to be a subterranean outlet from the Königs See, bursting out from a rocky cleft at the base of the Hohe Göll, at a point 1,916 ft. above the sea, or just 80 ft. lower than the Königs See. There are two principal falls, whose united height is 270 ft. The visitor should on no account omit to mount to the point where the stream is seen to burst out from the rock. Paths have been made to the best points of view, and there is a little inn supplying refreshments.

The paths from Golling to Berchtesgaden and the Königs See are noticed in Rte. F., and the Lammerthal, with the road to Abtenau, is described in § 46, Rte. E.

About 3 m. S. of Golling begins the remarkable defile called *Pass Lueg*, a very deep cleft between the Hagengebirge to the W. and the Tännengebirge to the E. The name is sometimes given to the entire valley as far as Werfen, but in a special sense to the narrowest part, at the N. end nearest to Golling. The road from that place keeps to the rt. bank of the Salza. About 2 m. from the village a finger-post is seen, with the direction 'Nach den Oefen der Salzache,' and the traveller will do well to turn aside and devote ½ hr. to following the winding path that will ultimately lead him back to the road at a point only ¼ m. above that where he left it. At this place a considerable bergfall seems to have flung huge masses of rock into the bed of the river, sufficient in some places to almost completely cover it over. The path is so well contrived as to lead visitors with no trouble to the most curious points of view. He peeps into vast caverns, formed by the blocks that have fallen together, and hears, though he can scarcely see it, the raging of the water that struggles below to escape from its temporary prison. A singular scene is sometimes beheld here, when the woodmen are let down by ropes into these dark recesses, in order to free the masses of timber, floated down from the upper part of the valley, that are often caught in the crevices of the rock. The path from the Oefen leads the visitor back to the road near to the chapel of Maria Brunneck (1,669'), whence he gains an excellent view of the opening of the defile of Lueg. At some points this is no wider than the exact space cut by the Salza, as it has gradually lowered the opening between the opposing rocks, and the road is artificially sustained on a ledge or shelf formed of pine trunks. Such a position, on so important a road, has naturally been much disputed in time of war. A little cavern above the road, now used as a fort, is called Croatenloch, from having been held by Croat troops in the war of 1742. In 1809, the Capuchin friar Haspinger, with a few Tyrolese volunteers, held the pass for a considerable time against a large force of French and Bavarians.

For several miles the defile continues wild and narrow, hemmed in between the high cliffs on either side. At the bridge of Aschau the road crosses to the l. bank, passing the solitary inn of Stegenwald, whose former host was one of the leaders in the struggle against the French invasion. From hence an excursion may be made to the Tännengebirge (§ 46, Rte. G). Before long the road passes the torrent from the *Blühnbachthal*, one of the finest of the alpine glens of this district. By that way the mountaineer may reach Berchtesgaden and Saalfelden (see Rte. F). The extensive smelting works, called Bluhhaus, at the opening of the glen, stand a little N of the Castle of *Hohenwerfen*, perched on a rock about 370 ft. above the level of the valley. This was the hunting-seat, stronghold, and state prison, of the prince-archbishops of Salzburg, some of whom are said to have rivalled the feudal chieftains, their cotemporaries, in the cruelties practised on the victims of their wrath. Passing over a sort of gap between the castle hill and the mountain on the W. side, the road reaches

Werfen (Inn : Post), in size a village, but ranking as a market town, 1,747 ft. above the sea, commanding a fine view of the Tännengebirge. The interesting ascent of the *Uebergossene Alp* may be undertaken from hence, by the Blühnbachthal, or by the parallel glen of the Höllenthal, which opens close to Werfen, or from the village of *Mühlbach* (where there is a fair country inn), about 7 m.

from Bischofshofen, connected with that place by a good road This great mountain, the highest limestone summit of this district, may be considered as a SE. prolongation of the range of the Steinerne Meer, which it resembles in its general structure ; but as the upper plateau is fully 1,000 ft. higher, the snow, which rests only in the hollows of the Steinerne Meer, here accumulates and forms a glacier of some extent. A cart-track through the Höllenthal, where iron ore has been extracted at many points, goes as far as *Höll* (3,154'), the highest hamlet. Above this fine pastures extend up to the Mitterfeld Alm (5,529'), where it is usual to pass the night. This point, whence the eastern face of the mountain rises very grandly, may be reached in 2½ hrs. from Mühlbach, passing the copper mine of Mitterberg. Few enjoy a greater variety of names. Besides those best known—Uebergossene Alp or Alm—it is often called Ewige Schnee, and sometimes Schneealm or Verwunschene Alm. The latter designation referring to a legend which frequently recurs in the Swiss and German Alps. To these must be added the name *Wetterwand*, by which it is known in Pinzgau. The highest summit, near the S. edge of the plateau, is (on the Werfen side) called *Hochkönig*, and measures 9,643 ft. above the sea The way ascends from Mitterfeld by the Gaisnase in 2 hrs. to the Ochsenriedel, a stony hollow at the base of a noble pinnacle of bare rock called Thorsäule. The remainder of the ascent, chiefly over a gently sloping and little crevassed glacier, takes 2 hrs. In 1865 the directors of the Mühlbach copper mine erected a small stone hut on the highest peak. There is room for two persons to lie down, but without firing the cold at night must be severe. It is possible, though rather difficult, to follow the ridge connecting this mountain with the Steinerne Meer. As a panorama the view from the Watzmann may be preferred, but the rock scenery here is much grander.

Above Werfen the valley of the Salza,

here called Pongau, opens out considerably, and from hence to St. Johann the mountains on either side are neither so high nor so bold in form, being in a geological sense, an eastern prolongation of the Kitzbühel and Dienten mountains. Some German writers have supposed that this portion of the valley was at some not very distant period a lake, similar in position to the Zeller See (Rte. B), which was partly drained by some convulsion that opened a passage for the Salza through the defile of Lueg. Few modern geologists will be inclined to accept this latter view. To establish the former existence of a lake, careful examination of the ground and accurate levelling will be required. Rather more than 1 m. above Werfen the post-road divides. One branch leading to Villach in Carinthia passes the river and goes through the Fritzthal, over a low col to Radstadt (§ 46, Rte. F). The road to Gastein and the Pinzgau keeps to the l. bank. Not far from the bridge, on the rt. bank of the Salza, is *Dorf Werfen*, also called Werfen Pfarr, not to be confounded with the Markt (market-town) of the same name. It has a fine old Gothic church, with a curious monument of Christoph von Künburg. The pedestrian should here turn aside, and ascend for some way by the Radstadt road, till he reaches a projecting point that commands an extremely fine view of the valley and the surrounding mountains, amongst which the Uebergossene Alp towers preeminent. If he would gain a still wider view, he may take the summit of the Gründeck (5,949′) in his way to St. Johann. On the road by the l. bank of the Salza is the very ancient village of *Bischofshofen* (Inn : Hirsch), containing two interesting churches. The more ancient is the Maximilians Kirche, founded by St. Rupert. The Frauenkirche is larger, but more modern. It contains some monuments, and several Roman antiquities found on the spot; e.g. a Roman altar, now used as a baptismal font. Near the village there is a fine waterfall of the *Gainfeldbach*, whence the bold peak of the Thorstein is seen

to the E. After passing the torrent from the *Mühlbachthal*, along which a new road is carried to the mining village of Mühlbach, mentioned above, and in Rte. G, the high road crosses the Salza, and soon after reaches St. Johann (Inn : Rosians, very fair). The village, called, for the sake of distinction, *St. Johann im Pongau*, stands on rising ground, about 390 ft. above the river, 2,034 ft. above the sea, a few miles below the point where (after flowing nearly due E. for 45 m.) it turns northward towards Salzburg. The road from hence to Radstadt, which mounts to the E. along the Kleinarlbach, is described in § 52, Rte. E. The road to Lend crosses the Salza, and is carried WSW. amid picturesque scenery, to the hamlet of *Schwarzach*. In the village inn is shown the identical table whereat the leaders of the Protestants entered, in 1731, into the covenant called Salzbund, whereby they bound themselves to quit their homes and country rather than surrender their religious convictions. The result of the persecution directed against them was the emigration of more than 22,000 persons from this part of the then principality of Salzburg. A short excursion may be made into the little glen of the *Wengerbach*, which opens on the high-road close to Schwarzach. At its head is a waterfall of great height, but with a small volume of water. On rising ground W. of the glen is *Goldegg* (2,686′), a village with an ancient castle, long inhabited by the powerful family who derived their name from this place. It contains a large hall, built in the 16th century, after the castle had passed from the Goldegg family to the Counts of Schernberg. The painted coats of arms of most of the princely and noble families of the time, accompanied by historical and allegorical details, will be likely to interest the antiquary. Above Schwarzach the valley of the Salza is contracted to a defile, not, however, comparable in point of grandeur to that of Lueg. The road recrosses the river to the rt. bank, and about 2 m. farther reaches *Lend* (Inns : Post; good), a straggling village extending along both banks of the Salza. Here the road to Mittersill and the head of the Pinzgau is carried due W. (§ 50, Rte. A), and that to Gastein mounts SE. through the Klamm (§ 52, Rte. A).

Route F.

BERCHTESGADEN TO THE VALLEY OF THE SALZA.

The traveller wishing to go from Berchtesgaden or Königssee to the valley of the Salza has a choice among many different routes, none of which is devoid of interest.

1. *By the Hangende Stein.* This is the only way fit for persons who travel in heavy carriages. They follow the road from Berchtesgaden to Salzburg, through the defile of Hangende Stein, to the point where the stream of the Albe turns eastward. A cross road, about 2 m. in length, keeps along the l. bank of the stream, and then joins the main road to Hallein, described in the last Rte. The distance this way is about 15 m.

2. *By the Dürnberg.* A road easily practicable for light carriages follows for about 10 m. a nearly direct line NE. from Berchtesgaden to Hallein, passing over the Dürnberg. In approaching Hallein it goes within ½ m. of the entrance to the salt mine (see last Rte.), so that the traveller may visit the mine and make his exit through the lower tunnel close to the town of Hallein with very little loss of time.

3. *By the Rossfeld.* The moderate walker, who prefers a mountain view to underground sight-seeing, may turn to the rt. from the last-mentioned road, and reach the summit of the *Rossfeld* (4,261′) by a pleasant and easy walk over alpine pastures. The summit is a northern outlier from the range of the Hohe Göll, and commands very pleasing views. It would be possible to combine this with a visit to the salt mine.

4. *By the Torrener Joch.* This extremely agreeable pass may be made from Berchtesgaden, or from the inn at Königssee, the time in either case being about 7 hrs. steady walking. In going from the Königs See it is well to remember that the true course keeps near to the Königsbach (Rte. A.) all the way, and the traveller should not be led aside by a tempting path to the rt. The ascent is tolerably steep, through a ravine with the Jenner on the l., and the Bärenwand on the rt., till the pastures of the Königsbergeralp are reached. There is here a gamekeeper's lodge. Due E. of this lies the *Torrener Joch* (5,697′), which may also be reached direct from Berchtesgaden by a less steep path than that by the Königsbach. The ridge falls but slightly on the E. side of the pass, and the track lies for ½ hr. over alpine pastures till it reaches the châlets of the Obere Jochalp, near to which the plateau ends abruptly, and a steep descent leads down to the *Blüntauthal*. On the Austrian side of the pass the E. edge of the plateau, which shows like the summit, is called Vordere Joch, and the Torrener Joch, or true summit, Hintere Joch. The view from the pass is not remarkable, but those gained during the ascent are beautiful, and the position of the Obere Jochalp is striking. From the pass the summit of the *Jenner* (Rte. A) may be reached in 40 min. To NE. is the *Hohe Göll* (8,266′), and beside it the *Hohe Brett* (7.690′), while S. of the pass rises the *Schneibstein* (7,424′). The Blüntauthal, through which lies the way to Golling, carries down to the Salza, opposite that place, the drainage of the range from the Hohe Göll to the Hagen-

gebirge. After descending nearly due E. for some distance the streamlet, along which the path runs, joins another coming from a southern branch of the glen, issuing from the heart of the Hagengebirge. Henceforward the way lies NE. until the path (passing near a fine waterfall) enters the valley of the Salza near the bridge opposite Golling. In taking the pass from that place it is important to follow the rt. hand branch of the Blüntauthal, where this joins that coming from the Hagengebirge.

5. *By the Blühnbachthal.* There is no doubt that the most interesting route for the mountaineer, going from Berchtesgaden or Königssee to Werfen, is by the *Blühnbachthal,* an extremely fine alpine glen that penetrates deeply into the midst of the highest peaks of this district; having the Uebergossene Alp to the S., and the chief summits of the Hagengebirge to the N. Its western end is closed by the ridge connecting the former mountain with the Steinerne Meer. The most convenient way to approach it from Königssee is to sleep at the Regenalm (Rte. A), and mount thence by the Landthal to the W. end of the range of the Hagengebirge. This may also be reached from the Obersee by a traveller starting from the inn at Königssee at early dawn. There are two passes by which the descent into the Blühnbachthal may be effected. The shortest course is by the *Blühnbachthörl* (6,609'), a depression between the *Klein Teufelshorn* (7,391') and the *Alpenriedhorn* (7,729').

Decidedly more interesting than this is the pass of the Wildthor, lying farther W., near the head of the Blünbachthal, right opposite and very near to the Uebergossene Alp. The track leading to it passes by a little tarn, called Blaue Lake, lying between the *Hocheck* (7,703') and the Ober Schönfeld. Passing a solitary châlet at the Vordere Wildalp, the track reaches the *Wildthor* (7,077'). A steep descent through a rocky cleft leads down to the head of the *Blünlachthal.* This fine glen, mentioned in Rte. E as opening a little

N. of the Castle of Hohenwerfen, is too rarely visited by travellers, but it is the favourite hunting ground of some neighbouring proprietors. From its head to the high-road is counted a walk of 4 hrs. About half-way is a handsome shooting-lodge (Jagdschloss) beside another old building designed for the same purpose by some former prince-bishop. Except when occupied by hunting parties, a stranger may obtain good accommodation here. The ascent of the Uebergossene Alp may be effected as well from the *Jagdschloss* (2,645') as from the Höllthal (Rte. E). Another interesting excursion is over the *Urschlauer Scharte* (6,889'), the ridge connecting the Uebergossene Alp with the Steinerne Meer. By that way Saalfelden is reached in 9 hrs.

ROUTE G.

SAALFELDEN TO LEND, BY THE URSCHLAUTHAL.

Reference was made in the introduction to this section to the ranges of triassic and palæozoic rocks lying S. of the high range of the Steinerne Meer and Uebergossene Alp, that are collectively known as the *Dienten Gebirge.* These are separated from the high limestone mountains by the *Urschlauthal,* or Urselauthal, mentioned in Rte. B as opening into the valley of the Saale at Saalfelden. By that way, and by the passes mentioned below, there are easy paths from Saalfelden to Lend, or to Werfen (Rte. F). That by Dienten is traversed by a good horse track.

The general direction of the Urschlau glen for nearly 3 hrs. is ESE., but the stream and the path wind considerably. 1¼ hr. from Saalfelden is the village of *Alm* (2,490'). Above this the glen contracts, and for more than 1 hr. continues in the same direction (under the name Vorderthal) to Mühllehen. Here the torrent makes a sharp turn, and the head of the glen, called Hinterthal, is seen opening to NNE. By that way goes the path to *Mosbach* (3,363'), the highest hamlet, very finely situated at the base of the precipitous rocks of the Uebergossene Alp. Even those who do not intend to traverse the Urschlauer Scharte, mentioned in the last Rte., will do well to visit the head of the Hinterthal. From *Mühllehen* the horse-track mounts ESE. to the *Filzen Pass* (3,953'), sometimes called Hochfilzen, but not to be confounded with the place so named on the way from St. Johann to Saalfelden (§ 44, Rte. F). From the summit the track descends to *Dienten* (3,041'), a small village possessing an ancient church, with pictures defaced by time, at the head of a short glen that descends a little E. of S. Iron mines were formerly worked here, and some interesting minerals are still found. About 10 min. above the village, near a building that is, or was, used for making iron nails, are rocks containing trilobites (?) orthoceratites, and other palæozoic fossils. ENE. of the village a track is carried over the *Dientner Alm* (4,517') into the *Mühlbachtal* which joins the valley of the Salza between St. Johann and Werfen (Rte. E). Instead of crossing the pass, the mountaineer should ascend the *Schneeberg* (6,292') lying a little to the S., due E. of Dienten, and may thence reach Bischofshofen (Rte. E) by the *Mühlbachthal*. The panorama is first-rate.

The torrent descending from Dienten preserves the ancient form of the name, being called *Tuonta*. This has cut an extremely deep ravine, and the track, after crossing to the l. bank some way below the village, winds round the shoulder of the mountain, commanding noble views of the high peaks to the S., and descends into the valley of the Salza a little above Lend. It is a walk of 4 hrs. from that place to Dienten. The track, which is practicable for carts, cannot be missed.

There is another way which is about equally short, but not to be so easily found. It avoids the Filzen Pass and Dienten, crossing the ridge some way W. of that village. The alp is called Grünberg, and the summit may be reached in 3½ hrs. from Saalfelden. In descending, the path to the rt. leads nearly due S. to Taxenbach. The way to Lend is either to follow the ridge on the W. side of the Dententhal, or else to descend into that glen at a point where it is possible to cross the Tuonta and follow the cart-track to Lend (Rte. E).

SECTION 46.

ISCHL DISTRICT.

In describing the road from Salzburg to Lend (§ 45, Rte. E) we had occasion to remark that the deep cleft in the Northern Alps marked by the valley of the Salza in that part of its course is very far from having the geological significance that belongs to the upper valley of the same river, or to the valley of the Inn between Landeck and Hall, or to several other valleys whose direction approaches due E. and W. At the defile of Lueg the river flows between the Hagengebirge on one side, and the Tännengebirge on the other, merely dividing into two portions what was once a continuous mass; and there can be little doubt of the orographic connection between this and the still more considerable mass of the Dachsteingebirge, whose axis lies in the prolongation of the first-named range. Northward from this principal mass are a series of outlying ridges and isolated groups, distributed with little apparent order. In the midst of these outer ranges lie the numerous lakes that have made the district here described justly celebrated for the variety and beauty of its scenery. In the centre, and at a convenient distance from several of the most picturesque lakes, is Ischl, which within a comparatively short time has risen to importance as one of the most attractive of Alpine watering-places. The small tract surrounding that village, extending to the head of the Lake of Hallstadt, has long been of importance for the salt-mines that have rivalled those of the somewhat similar district of Reichenhall and Berchtesgaden, and on this account has long formed (under the name *Salzkammergut*) a distinct dependancy of the Austrian crown.

The Alpine district, however, included in the present section extends considerably beyond the limits of the Salzkammergut; and it has therefore seemed better to take the name from Ischl, now universally known as a centre of attraction for travellers visiting the Eastern Alps. The western boundary is fixed by the valley of the Salza, and that to the S. by the road from Werfen to Radstadt, and thence along the valley of the Enns to Steinach. To fix the eastern limit is not so easy, but the most natural boundary seems to be the valley of the Traun as far as the N. end of the Lake of Hallstadt, and thenceforward the line of depression followed by the Aussee road, entering the valley of the Enns a little way above Steinach.

The detached masses of the Traunstein, the Schafberg, and Höllgebirg, are the northernmost notable summits of the Alps, all of them lying north of any part of the Swiss territory. Although the highest summits of the Dachstein group—Hoher Dachstein (9,845'), Thorstein (9,677'),—do not reach 10,000 feet, there is a greater extent of glacier than would be expected by a traveller familiar only with the central and southern ranges of the Alps; and these and the associated peaks offer to the enterprising mountaineer the attraction of difficulty which in regard to some of them has not yet been surmounted.

The traveller finds tolerable quarters everywhere in the inhabited places in this district, and comfortable inns at many of the more attractive stopping-places. It is to be regretted that mountain inns such as those now common in Switzerland have not yet been provided to meet the wants of travellers. This deficiency is particularly felt at the Vorder Gosausee, one of the spots most attractive to the mountaineer that can be named in the Eastern Alps.

The visitor who would become thoroughly acquainted with the charming scenery of the outer ranges of this district, near Salzburg, should consult two papers, by Dr. Wallmann, in the *Jahrbuch* of the Austrian Alpine Club, vols. ii. and iii.

ROUTE A.

SALZBURG TO ISCHL. EXCURSIONS FROM ISCHL.

	Austrian miles	Eng. miles
Hof	2	9¼
St. Gilgen	2	9¾
Ischl	3¼	15¼
	7¼	34¼

Diligences ply daily between Salzburg and Ischl, employing 6¾ hrs.—fare about 4 flor. There is also an omnibus which takes 1 hr. longer, at a fare less by 1 flor. For hired carriages (lohnkutscher) the charge is 8½ flor. with one horse, 12½ flor. with two horses, with 1 flor. trinkgeld to the driver,—time about 9 hrs.

The great majority of travellers who take this road make Ischl the object of their journey, intending to remain many days, or even weeks, at that place, and making it the centre for the numerous excursions there offered to the choice of the stranger. But it may also be very well taken on the way from Salzburg to Vienna or to Trieste. The road from Ischl to Steinach (Rte. D), and thence to Bruck an der Mur on the rly. from Vienna to Trieste (in all, about 108 Eng. m.), is throughout very interesting and may easily be accomplished in two days.

The first stage from Salzburg is almost constantly up-hill, so that the pedestrian gains little time by taking a carriage. Leaving Salzburg (§ 45, Rte. A) by the Linzer Thor, the road passes close under the Nockstein, a promontory from the Gaisberg. Looking back, the traveller gains a wide view over the plain of Bavaria and the mountains beyond Salzburg. Soon after reaching the level of an undulating and broken plateau, a road to Ebenau turns off to the rt., while the high-road keeps on nearly due E. to

Hof (2,203′), where the omnibus stops, or did stop, for dinner at a bad inn. The place consists of this, the church, and two or three houses. The complaints against the inn and the food here have been so general that travellers were advised to dine either before or after the journey, or else take refreshments with them; but it is fair to say that some recent accounts are less unfavourable. Travellers from Gastein or the upper valley of the Salza to Ischl usually follow the road from Golling by Abtenau (Rte. E), but those going from Berchtesgaden to Ischl, who have no occasion to go to Salzburg, will do well to reach Hallein by any of the roads or paths mentioned in § 45, Rte F., and then follow a country road to Ebenau,

an easy walk of 3½ hrs. from Hallein, and 1 hr. more takes the traveller to Hof. This road from Hallein to Hof is passable for a country carriage. In going from Hof to St. Gilgen the traveller not pressed for time should make a slight détour by the Mond See, unless he intends to make the latter the aim of a special excursion (see Rte. C).

From Hof to Ischl the road presents a succession of charming pictures, forming a fit introduction to the fascinating scenery of this district. The first lake seen is the *Fuschelsee* (2,097′), a narrow basin of dark-blue water, with an old castle at its NW. end. The road runs along its S. shore, passing at the eastern end the little village of *Fuschel* (Inn: Mohr), and then mounts the low ridge that divides this from the next lake. The view looking back over the Fuschelsee is very pleasing, but no way comparable to that which is gained when, after passing the col (2,525′), the road begins to descend the SE. slope towards the Wolfgang See. The poverty of language, and the need of economising space, counsel the writer to avoid the frequent use of epithets. Suffice it to say that the traveller who, with moderately favourable weather, visits this and the other lakes of this district without exquisite enjoyment, must be devoid of all sense of the beauty of nature. In descending from the pass, the whole length of the lake is seen stretching to SSE. for about 7 m. with richly wooded slopes of moderate steepness. above the rt. or S. shore, contrasted on the opposite side with the extremely bold range of the Schafberg. The ridges and promontories that stretch out from it towards the lake seem to have been cut away as they approach the water, and one of them in particular—the *Falkenstein*—shows a vertical precipice of pale grey limestone, that harmonises in a marvellous way with the dark green of the pine woods, and the azure blue of the lake.

The lake, and the ascent of the Schafberg, are described lower down among the excursions from Ischl, but many travellers find it a better plan to halt on their way to Ischl at the village of St. Gilgen (Inn: Post, tolerable, not cheap), whence the ascent is as short and interesting as that from St. Wolfgang. Those who do not require a carriage to convey their luggage will do well to engage a boat from St. Gilgen to Strobl, a hamlet at the E. end of the lake, about 7 m. from Ischl.

The post-road is carried along the S. shore, where a promontory stretches out and nearly divides it in two parts. This has been formed from the débris deposited by the *Zinkenbach*, a stream descending from the *Hoher Zinken* (5,778'). A path here turns SSW., and follows the torrent to the fine waterfall of the *Schreinbach*, about 1 hr. distant from the road. From Strobl the road follows the Ischl, which is the outlet of the lake, crossing the stream, and again returning to the rt. bank. There is a foot-path by the rt. bank nearly 1 m. shorter than the road.

ISCHL. (Hotels: A very large and handsome new house managed by Herr Bauer was opened in 1865—beautifully situated, but the charges extravagant; Kaiserinn Elizabeth, burnt down in 1865, reopened in 1868; Kreuz, very well kept, not dear; Post, tolerably good; the following are the best among the second-class inns— Erzherzog Franz Karl; Stern; Baierischer Hof) stands close to the junction of the Ischl with the Traun, 1,595 ft. above the sea. Though the most populous place in the Salzkammergut, and styled a market town, this was 25 years ago no more than an Alpine village, but its almost unequalled advantages of position, and the reputed good effects of its baths, had already begun to attract visitors. Among these were the late Emperor Ferdinand and his Empress, who often resided in an ordinary house in the village. The continued preference shown to the place by the imperial family, and by a large portion of the wealthier classes in Austria, has made it a sort of Alpine capital, where, during two or three summer months, the leaders of fashion and ministers of state combine with their ordinary occupations something of the enjoyments of mountain air, exercise, and beautiful scenery. The Emperor's villa is as poor a specimen of architectural art as the garden is attractive by the taste with which it is laid out, and the admirable view which it commands. Several Austrian noblemen have also built villas close to Ischl, and several buildings of a public nature contribute to give to modern Ischl the air of a provincial capital. Among these may be reckoned the baths and Trinkhalle (a large building in the Grecian style), a theatre, a casino, a good circulating library, cafés, &c. It naturally follows from the altered condition of the place that Ischl is by no means a cheap summer residence; yet, as compared with places of a similar character in England or France, it cannot be called very expensive. Those who remain for some time will find it economical to engage lodging in a private house. Except at the most crowded season, these are found without difficulty. A stranger remaining over 6 days pays what is called a *Kurtaxe* of 5 fls., with 1 fl. more for each additional member of the same family. The produce of this tax, which must annually amount to a large sum, is entrusted to an Improvement Committee (Verschönerungscomité). It is complained that the money is mainly applied to what we should call parochial purposes, and that little or nothing is now done for the benefit of the strangers from whom it is raised. There are, indeed, numerous walks well laid out and planted, with benches, and monuments of various degrees of artistic merit, close to Ischl, but these have existed for many years; and it is justly said that the funds should be applied so as to facilitate the excursions which are the main attractions of the place, by improving mountain paths, and cross-roads, &c.

The curative means employed here, under the direction of one or other of the competent physicians who reside here in summer, are even more varied than at Reichenhall. The brine is employed for baths, and the steam from the evaporating pans is used either for vapour baths, or inhaled by patients who walk about in a covered gallery over the pans. Mud from the salt mine is also applied externally in certain cases. In addition to these there is a sulphureous spring used for baths, either by itself, or in combination with the brine; and the whey-cure (molkenkur) is followed by other patients, with or without the use of baths.

Of short walks about Ischl a long list might be given, but this is scarcely necessary. Among the most agreeable are, those to the ruined Castle of *Wildenst in* 1 hr. from Ischl, to the Rettenbach Mühle, rather less than 1 hr. distant, and thence to the so-called Wildniss, about 1 m. farther; and lastly, that to the Hohenzoller waterfall, returning through the Jainzer-thal.

Of the longer excursions from Ischl, many are commonly made partly or altogether by carriage. These are usually to be found standing ready for hire, but in the full season those who would avoid disappointment should secure their vehicle beforehand. In making an agreement for a carriage, it should be stated that the price includes extra horses (Vorspann), which are required on some mountain roads.

The tariff established by the local authorities is hung up in all the hotels. The rate is about two fl. per Austrian mile for those with 1 horse, and about 3 fl. for those with 2 horses—Trinkgeld, 50 kr. for half a day; 1 fl. for the entire day. The charge is the same whether the carriage be kept for returning to Ischl or not. Persons requiring posthorses must give 2 hrs. notice to the postmaster. The best guides for mountain excursions are Schütz, Graf, Karl Neff, Hütter Flörl, and Richer. The pay for a day's walk is 2 fl. The charge for a chaise-à-porteur (Tragsessel) is fixed by tariff for the excursions usually made by ladies, being about 5 fl. for the shorter excursions, and nearly double as much for those occupying the whole day. Each bearer expects in addition a trinkgeld varying from 20 to 55 kr. The charge for excursions not enumerated in the tariff must be fixed by agreement.

We now proceed to notice the more interesting of the excursions from Ischl, remarking that, according to the route selected by each traveller, he may take one or more on the way to and from that place.

1. *To the Wolfgang See, and ascent of the Schafberg.*—The *Wolfgang See*, more properly called St. Wolfgangs See, and also Aber-See, has been already mentioned as lying on the N. side of the road from St. Gilgen to Ischl. Visitors from Ischl may take the carriage to St. Wolfgang, or stop at Strobl, and there hire a boat. This lake attracts travellers as well by the extreme beauty of its scenery, as because it is the more convenient way for approaching a mountain that commands one of the most perfect panoramic views to be found in the entire range of the Alps. From whichever side it be approached, either by St. Gilgen in coming from Salzburg, or by Strobl at the end nearest Ischl, no traveller should omit to traverse the lake in a boat, whence alone the scenery of its shores can be fully enjoyed. The utmost length is not more than 7 miles, and its height above the sea 1,751 ft. The greatest depth is said to be 360 ft. Without rivalling in grandeur the Königs See, or the Hallstädtersee, or the Bay of Uri in the Lake of Lucerne, it may be compared for picturesque beauty with any one of the smaller lakes of the Alps. The feature which best serves to fix its peculiar physiognomy on the memory is the grand precipice of the *Falkenstein*. On arriving opposite the great mass of grey rock that rises from the water's edge, the boatmen are careful to exhibit the remarkable echo by shouting out 'Heiliger Vater Wolfgang, komm ich zurück, sag ja.' In ordinary fine weather the echo answers 6 or 7 times 'ja;' but when the air is troubled the answer is faint, or scarcely heard at all. As this often happens when bad weather is impending, the question is more appo-

site than might be supposed. The chief village on the lake shore, and that best deserving a visit, is

St. *Wolfgang* (Inns: Rössel, tolerably good; Zum Cortisen, complaints of the attendance; Hirsch, said to be dear; this and the last enjoy a fine view of the lake), standing on the narrow space between the foot of the mountain and the lake. There is a curious Gothic church here (built in the earlier half of the fifteenth century) containing the shrine of St Wolfgang, and several remarkable specimens of early German art. Especially curious is the high altar, with a central compartment carved and coloured by Pacher of Brunecken, in or about 1481. The wings contain paintings on panel attributed to Wohlgemuth. The original chapel built on the rock by the saint is seen in the middle of the church. Opposite the building is a curious fountain, dated 1515, with a bronze statue of St. Wolfgang, standing on a pedestal whence issue four jets of delicious water.

The main object of most strangers visiting St. Wolfgang is the ascent of the *S hafberg* (5,837'). Of the many mountains throughout the range of the Alps that have been compared to the Rigi, this alone presents a tolerably close resemblance, for the reason that here the most varied and charming effects of lake scenery are combined with a noble Alpine background. Three lakes—the Wolfgangsee, Mondsee, and Attersee—approach close to the base of the mountain; and beyond these the eye rests on not less than nine or ten others, extending in the far W. to the Chiemsee and Simmsee, not to speak of the many smaller sheets of water that lie imbedded, like blocks of sapphire, in a setting of grey rock and emerald Alpine pasture. The Alpine panorama extends from the range of the Todtes Gebirg to the Steinberg and Lofer Alps, with a peep at the distant Tauern range; but the objects that engross most attention are the grand peaks of the Dachstein group, whose keen shafts break through the coating of glacier that covers their middle height, and point upward to the sky.

The landlord of the Rössel at St. Wolfgang is the owner of a mountain inn on the top of the Schafberg, recently enlarged, and made very comfortable. During the crowded season beds can be secured only on presenting a ticket from the landlord at St. Wolfgang, so that most travellers start from that place. Practised mountaineers do not require a guide, unless it be to carry a knapsack. The charge is 1½ fl., or 2 fl. if the descent be made by St. Gilgen; mules cost 9 fl.; and Tragsessel 10 fl., with a trinkgeld of 25 kr. to each bearer. For staying the night at the top, each man is entitled to half a florin additional. A slight addition to these charges is made if the traveller descend on the N. side of the mountain to Schärfling or Unterach. One of the guides here, named Panzner, is well acquainted with the entire district, and has picked up some knowledge of the local geology from the Austrian professors whom he has accompanied. He is said to be also a good mountaineer. The beaten track is very rough, but cannot easily be missed. In 2½ hrs. the traveller, going gently, and stopping to admire the charming views, may reach the Oberalp, where he joins the track from St. Gilgen. There is here a little inn belonging to the master of the post at that place; but when that at the top is not over full, it is much to be preferred for passing the night, as at least 1 hr. of a rough ascent is required to reach it from the Oberalp.

The mountaineer who does not object to lengthen somewhat so easy a day's walk, is advised to take the top of the Falkenstein on his way from St. Wolfgang to the Schafberg. There are few points whence it is possible to look down so directly upon an Alpine lake, and the effect is very beautiful. A guide is needed to find the way from the Falkenstein to the Schafberg.

The latter mountain is ascended very frequently from St. Gilgen, and it is a good plan to vary the route in going or returning.

The N. side of the Schafberg is much steeper than that facing the Wolfgangsee, in some places forming vertical precipices; but there is no difficulty in descending by the paths leading to Scharfling on the Mondsee, or to Unterach on the Attersee, places which have much the same relative positions as Immensee and Kussnacht in respect to the Rigi.

The geologist who has mastered the somewhat intricate relations of the secondary strata in the mass of the Dachstein group, will find the same beds repeated in the Schafberg, which is in truth only the highest part of a ridge extending more than 20 m. from the Schober, near Mondsee, to Ischl, and is perhaps orographically a western extension of the main range of the Todtes Gebirg, described in the next section.

The northern botanist will be interested by finding many Alpine plants on the Schafberg; but, as compared with several other mountains of this district, it cannot be called productive.

2. *To the Gmundensee, and Falls of the Traun.*—This expedition should not be omitted by those who do not approach Ischl by the Linz road, in connexion with which (Rte. B.) they are described in this work. In going from Ischl, the most agreeable way is by boats on the Traun. The operation of shooting rapids is scarcely to be performed elsewhere in Europe so safely and agreeably. In that way the Lake of Gmunden is reached in less time than by carriage.

3. *To the Attersee and Mondsee.*—The way from Salzburg to Gmunden, passing both these lakes, is described in Rte. C. Those who reach Ischl from Linz, or from the south, may make a very interesting excursion, requiring at least three days, in which they may visit both these lakes and return by the Schafberg and Wolfgangsee. They may go from Ischl to Weissenbach, at the SE. corner of the Attersee, by a good char-road over the low col dividing the Ziemitz from the Höllgebirge. This is in itself a very agreeable drive, giving a charming view of the Attersee, and is worth making from Ischl by those who do not intend to make the entire round.

Another way to Weissenbach, practicable only on foot, is by Wirer's Waterfall, reached by a path turning to rt. from the road to St. Wolfgang. Above the pretty fall a path bearing to l., or westward, leads to the lonely little *Schwarzensee*, a small sheet of water about 1 m. long and ½ m. broad. Above the lake the traveller continues to ascend northward, amid Alpine pastures, for 2 hrs., until he suddenly finds himself on the verge of the very steep face of the mountain overlooking the Attersee and Mondsee. A good path leads down to Weissenbach (Rte C), whence he may reach Mondsee on the same day. In returning he will follow the carriage road from Schärfling to St. Gilgen, a distance of only 4 m., passing the *Krötensee* (1,880'), a picturesque little tarn lying in a hollow between the *Drachenstein* to W., and the Schafberg to E. In this solitary spot Prince Wrede has built himself a modern castle in the style of Hohenschwangau (§ 42, Rte. A.) The ascent of the Schafberg may be made from this point, and the traveller may return to Ischl by St. Wolfgang.

The path to the Schwarzensee mentioned above is often taken by pedestrians going from St. Wolfgang to Ischl, who return to the road between those places by a remarkable cleft in the rocks, through which the Schwarzenbach torrent descends in a fine waterfall called *Wirer's Strub*. The descent along the face of the rock by a path overlooking the chasm is scarcely advisable for nervous persons, though the sesselträger from Ischl sometimes carry ladies that way.

4. *To the Aussee and Grundelsee.*—These two charming little lakes were too long neglected by visitors to Ischl, but of late years are much frequented, and the good inns there are often filled by persons who remain weeks, or even months, to enjoy the many advantages of the neighbourhood. The post-road to the little market town of Aussee is

described in Rte. D., but the lake from which it takes its name is nearly 4 m. distant, beside the much smaller village of Alt-Aussee. In making the excursion from Ischl, the pedestrian should approach the lake through the glen of *Rettenbach*, which opens into the Traun valley about ½ m. below that place. As far as the above-mentioned Wildniss, this is a favourite stroll for visitors. In 3 hrs. from Ischl the traveller reaches the Rettenbach Alp. To the rt. he has the ridge of the *Sandling* (5,619'), to the l. the *Hohe Schrot* (5,691'), a western promontory from the mass of the Todtes Gebirg. The track, which is sufficiently marked to make a guide almost superfluous, soon turns southward, traverses the pass between the Sandling and the Loser (2½ hrs. from Ischl), descends by Ramsau along the Augsbach to *Fischerdorf* on the lake shore, and thence along its margin to *Alt-Aussee*, where there is a good inn, more attractive than those at Aussee.

The *Aussee* (a corruption from Augsee), though only 2 m. long, and ¾ m. broad, is one of the gems of this district. The scenery is of the highest order, especially from spots above its N. shore, where the grand mass of the Dachstein peaks is seen in the background. Of nearer objects the most striking are the *Loser* (5,799'), rising due N., and the *Trisselwand* (5,865') E. of the lake. The lake is 2,247 ft. above the sea, and, like all those of this region, abounds in delicious fish. The ascent of the Loser is an expedition much recommended. Practised mountaineers wishing to see something of the Todtes Gebirg (§ 47, Rte. C.) may traverse very rough ground between the Loser and a ridge called Klopf, whence they may descend on the l. hand to the Augstwiesenalm on the plateau of the Todtes Gebirg, or on the opposite side to the Aussee.

From Alt-Aussee those who have not yet seen any of the salt mines of this region may visit the Ausseer Salzberg, 1 hr. distant. The general plan of working is the same as at Hallein, but this is more interesting to the mineralogist from the variety of other substances here found associated with the common culinary salt. For sight-seers the Ischl Salzberg (Excursion 7) is more interesting.

To visit the Grundelsee the traveller may follow the rough road from Alt-Aussee to Aussee, about 4 m. At first it keeps away from the narrow cleft through which the Ausseer Traun descends from the lake, but lower down runs beside the stream to that village (Rte. D.), whence a pretty good road is carried to the Grundelsee, a distance of nearly 3 m. A shorter way, and one commanding finer views, is by a path leading across the low ridge of the Tressenstein direct from Alt-Aussee to the *Grundelsee* (2,164'). It is not easy to give by description an idea of the special attractions of this lake. It is enough to say that every one who has visited it counts it among the most charming scenes to be found in this district, so wonderfully rich in natural beauty. At the house of the Fischmeister, at the W. end of the lake, strangers find clean and comfortable quarters; and there is another inn on the shore at a spot called Ladner. The best course is to engage a boat to the farther end of the lake, about 4½ m., and then, following the valley eastward, to visit two smaller lakes, the *Töplitz See* and *Kammer See*, the latter lying at the foot of the *Weisswand*. The Grundelsee is one of the best centres for excursions into the range of the Todtes Gebirg, further noticed in § 47, Rte. C.

Whether the traveller intend to unite the visit to these lakes with the excursion to Hallstadt, next described, or merely to return to Ischl, he should on no account omit the beautiful valley leading from Aussee to *Obertraun*, at the head of the lake of Hallstadt. Less than 1 m. below Aussee the united torrents, called respectively *Ausseer Traun* and *Grundelseer Traun*, receive a third affluent from SE., called

Oedenseer Traun, and the stream, henceforward bearing merely the name

Traun, follows a somewhat tortuous course through a cleft between the S. end of the range of the Sarstein and the Koppen. A char-road is carried through the valley down to the village of *Ober-Traun*, but the exquisite scenery will be better enjoyed on foot. It is a walk of only 2½ hrs. On the way the pedestrian should turn aside and visit a curious cavern called *Koppenbrüllerhöhle*, through which a torrent rushes with a strange sound, that has given its name to the spot. This has lately been made more accessible, and it may be easily reached from Hallstadt, the excursion requiring only 4 hrs. In the lower part of its course the Traun flows through a nearly level valley, rich with noble sycamores, between grand ranges of limestone rocks. The still mightier masses that rise above Hallstadt, on the opposite side of the lake, close the view, which for grandeur and beauty may vie with the most famous scenes in the Alps. Taking a boat at Ober-Traun, Hallstadt is reached in 20 minutes; or if bound for Ischl, the traveller may go direct to Steg in 1½ hour. For the ascent of the Sarstein, which may be combined with this excursion, see Rte. D.

5. *To the Hallstädter See.* Those visitors who content themselves with a single excursion from Ischl usually select that to the lake of Hallstadt, and probably with justice. It does not combine the softer with the sublimer beauties of alpine scenery so completely as some others, but in several respects it must be counted as among the most remarkable of alpine lakes. There is none other of any importance that lies so immediately under a great mass of ice-clad peaks, and there are few that are looked down upon by so bold and stern a range of precipices as those that encompass nearly the entire basin.

The lake is a little over 5 m. in length, near 1½ m. in breadth, and the greatest depth is said to be 415 ft. Considering that we are here at the very foot of a mass whose summits approach the limit of 10,000′, the height above the sea level (1,769′) is very slight, being in fact less than that of a great part of the Bavarian plain. On the E. side the lake is walled in by the ridge of the *Sarstein* (6,558′), an almost unbroken range, presenting a steep curtain of pine forest surmounted by a continuous face of limestone rock. To the W. and S. rise the far grander and more broken masses of the Dachstein Gebirge. The two highest peaks lie at some distance from the lake, but the impending summits are high enough to give a character of grandeur to the scenery without altogether shutting out the view of the glaciers that encompass the higher summits.

The way from Ischl is by the post-road along the valley of the Traun until, a short way beyond Goisern (Rte. D), a road turns off to the l., and about 9 m. from Ischl reaches *Steg*, a hamlet on the lake shore close to the Traun. It has a tolerable little inn. It is impossible not to compare the position of this place with that of the similar hamlet of Königssee, on the lake of that name. The similarity is in many respects very striking. If this be allowed to surpass its rival in stern grandeur, the other may claim the palm for variety.

A very small steamer now plies on the lake of Hallstadt, but many travellers prefer to take a boat, which is obtained for a trifling hire. It is indeed possible to follow the carriage road for some way along the W. shore as far as the Gosaumühle, at the opening of the Gosauthal (Rte. E), but the scenery is far better seen from the lake.

A fine view of the lake and the valley of Gosau is gained from a point on the lower slope of the Sarstein which is reached by a path constructed at the expense of Professor Simony of Vienna, one of the most successful explorers of this district. Rather more than 1 hr. is necessary to go by boat from Steg to

Hallstadt (Inns: Beim Seeauer; Post; both belonging to the same proprietor, and both very well kept; Grüner Baum, very fair). This singular village stands on the narrow space left between the base of the mountains and the water's edge. There is absolutely no level ground available, and the houses are built on successive ledges of rock, communicating by stone steps. It is not quite correct to say that it is accessible only by water, for in both directions practicable paths are carried along the lake; but they are little more than goat tracks, and that leading to the Gosau valley mounts more than 1,000 ft. above the level of the lake. The inconsiderable Mühlbach torrent falls through the village, and works two or three mills. The church is ancient and curious, dating from 1320 ; it stands on a terrace above the village, which affords a charming view of the lake. Here, as elsewhere in the Salzkammergut, a large portion of the population are Protestants, and a new church has been constructed for their use.

The more interesting spots commonly visited by those who make the excursion from Ischl are here noticed, but the more laborious expeditions amidst the upper peaks of the Dachstein group will be described in connexion with Rte. E.

An ascent of 50 min. by a good path takes the visitor to the *Rudolfsthurm*, built by the Emperor Albrecht in 1284 for defence against the attacks of the Archbishops of Salzburg, and named by him after his father, Rodolph of Habsburg. An inscription by the way records a visit made to the neighbouring mine by Kaiser Maximilian in 1504. The tower, which is 1,150 ft. above the lake, is now used as the residence of the intelligent director of the salt mine. He has placed here a collection of minerals and fossils, and a portion of the remarkable objects of antiquity found in the so-called Celtic burying-places near at hand. Various articles worked in winter by the miners out of the marble of this locality are here kept for sale. The views from the windows of the tower are most interesting. The ascent to this spot is well worth making, even by those who do not visit the mine, or follow the path leading from hence to the Gosauzwang (about 2 hrs. distant), where the brine pipes from the salt mine are carred on arches across the opening of the Gosauthal, near the spot where the carriage road turns away from the lake to enter that valley.

The salt mine (Hallstädter Salzberg) is excavated into the triassic strata of the mountain above Rodolph's tower. The chief entrance is about 500 ft. above it, and it is easy to obtain admission; but the traveller in this district has other opportunities for visiting similar mines, which are all worked in the same manner as that of Hallein (§ 45, Rte. E) or Berchtesgaden.

Near to Rodolph's tower, in a little depression between the Siegkogl and the Kreuzberg, a large number of ancient graves have been uncovered during the last twenty years—in all over 850—and a quantity of arms, implements, and ornaments, chiefly of bronze, have been extracted. The more important objects were removed to the Imperial Museum of Antiquities in Vienna, but some are to be seen in the tower. The best account of them has been published by Professor Simony (Vienna, 1851). In conformity with prevailing ideas, they were at first considered Celtic. According to the present state of knowledge they are referred to the latter part of the bronze age and the beginning of the iron period.

Of the shorter walks near Hallstadt there is none so interesting as that to the *Waldbach-Strub*. The way to it lies through a short and beautiful glen opening close to the south end of the village. After making his way up and down several flights of steps amid the houses, the traveller sees the foaming torrent that descends from the waterfall. He follows its course between massive walls of rocks, whose ledges are fringed with trees, by a path winding amid huge blocks fallen from the heights above. After half an hour of easy walking, he reaches the head of this recess in the mountains, and sees a track mounting to the right towards a dark cleft in the mountain. The deep tones of the fall, and the cloud of spray seen through the trees, would alone suffice to guide a stranger to the principal fall, which is one of the most picturesque in this land of waterfalls. Several paths have been carried to the most favourable points of view, and the excursion is easily accomplished by most ladies. Two hrs. (going and returning) are sufficient. There is another fall, by some considered the finer of the two, reached in one hour's ascent from the lower fall (see Rte. E).

The rough track, leading from Hallstadt round the S. end of the lake may be followed by those who would visit the two curious spots called Hirschbrunn and Kessel, or these may be visited with less trouble by taking a boat, and landing at the nearest point of the shore.

The *Hirschbrunn* often exhibits nothing but a mass of waterworn rocks and pebbles, but after heavy rain, or in hot weather when the Dachstein snows are melting fast, a considerable volume of water here breaks out at a point a few feet above the level of the lake. The *Kessel* is a hollow on the slope of the mountain wherein, as in a caldron, lies a small pool of water. Under the circumstances mentioned above, this is rapidly filled to the brim, and the water bursts out with a gurgling roar and rushes down to the lake. Those who visit this part of the shore, whether on foot or by boat, should not fail to extend the excursion to Ober-Traun (see Excursion 4). They may then take boat back to Hallstadt or to Steg. In returning from Steg to Ischl, travellers often avail themselves of the barges that carry salt and other stores down the river, an agreeable and rapid mode of conveyance. Though there are several rapids, where the boats are hurried down between projecting rocks, accidents are unknown. In the dry season, the water of the lake is held back by massive sluices, in order to provide a sufficient volume when these are opened, to carry the boats down to Ischl and to the lake of Gmunden.

The neighbourhood of Hallstadt and Gosau is classic ground to Austrian geologists; owing to the comparative abundance of fossils, the relations of the triassic and jurassic strata have been pretty completely made out, and have served as a key to the right understanding of the other sedimentary ranges of the Eastern Alps. A residence of some days at both the above-named places must be as interesting to the geologist as to the lover of grand natural scenery. Wallner (see Rte. E), would be a valuable guide to the geological traveller.

6. *To the Gosau Lakes.* For the excursion to Gosau the reader is referred to Rte. E. Even though the traveller should go no farther than the Vorder-Gosausee, this is one of the most interesting expeditions from Ischl. To reach the Hinter-See, which lies in the very midst of the highest peaks and glaciers of the Dachstein group, is just possible in one long day, going from and returning to Ischl; but it is far more advisable to sleep at least one night at Hinter Gosau, though the accommodation is rough.

Precedence has been given to the lakes, as they undoubtedly offer the strongest attractions to the lover of nature in this part of the Alps. We now enumerate some other expeditions that may conveniently be made in one day from Ischl.

7. *To the Salzberg.* The salt mine near Ischl, called for distinction *Ischler Salzberg*, will not much attract those who have already seen those at Berchtesgaden or Hallein. Other travellers may be tempted to make the excursion, which has at least the attraction of novelty. The system of working is the same as that described in the Hallein mine, but there is the advantage here, that on certain days, of which previous notice is given at the hotels in Ischl, the galleries are lighted up; but in the mines of this region the salt is intermixed with earthy particles, and the surface does not exhibit any trace of the crystalline nature of the mineral. Having previously obtained tickets of admission at the head office in Ischl, visitors go by carriage road for about 3 m. to the hamlet of *Perneck*, or Berneck. Thence the ascent is made on

foot or by tragsessel, to the main opening of the mine, 3,170 ft. above the sea. The mine is not open to visitors on Saturdays or Sundays.

8. *To the Ziemitz.* It has been already remarked, that the Schafberg is but the central and highest portion of a range extending from the Schober, near Mondsee, to Ischl. The *Ziemitz* is the name given to the eastern end of this range overlooking the latter village. Its highest point, called *Leonsbergzinken* (5,071'), is reached in 4 hrs. from Ischl, and commands a very fine view of the neighbouring valleys, with three lakes, and with the Dachstein group in the back-ground.

9. *To the Hohe Schrot* (5,691'). Mention has already been made (Excursion 4) of this mountain, lying in the angle between the Rettenbach and the Lower Traun, and forming the WNW. extremity of the Todtes Gebirg, unless we prefer to consider the ridge, above spoken of, from the Ziemitz to the Schober as a continuation of the same range, separated only by the cleft through which the Traun descends to the plain. The ascent of the Hohe Schrot is longer and rather more laborious than that of the Ziemitz, but the view is finer, including the greater part of the Lake of Gmunden, and part of that of Hallstadt, with a noble alpine panorama.

10. *To the Kahlenberg* (6,066'). Visitors to Ischl are often induced to ascend the slopes SW. of the village, for the sake of the views gained over the valleys of the Upper Traun and the Ischl. The mountain lying in the angle between these valleys is the *Katerberg* (4,758'), and 1 hr. farther is the somewhat higher summit of the *Hainzen* (5,364'). These form portions of a ridge extending NE. from the mountain mass lying S. of the Gosau valley, and this higher intervening range cuts off the best part of the view from the above-named summits. The highest summit of this group (sometimes known as the Ramsauer Gebirge) is the *Kahlenberg*, and the ascent is but little more fatiguing than that of the Hainzen, and much more interesting, as it commands an admirable view of the Dachstein peaks and glaciers, which are not more than 7 or 8 m. distant. The ascent is made without difficulty from Goisern, but still more easily from Gosau.

11. *To the Chorinskyklause.* Mention has been made, in § 43, of the system of dams generally used in the Eastern Alps for barring the course of a stream, and using the pent up water, set free by a sluice-gate, for carrying timber down to a lower valley. The very large consumption of timber incidental to the extraction of salt from brine has led to the construction of these barriers—called Klause—at many places in the mountains near Ischl. The largest of these, called Chorinsky klause, is designed to bar the stream of the Oberweissenbach, which descends from the W. into the valley of the Traun near Lauffen, a village on the road to Aussee, 3 m. above Ischl. A walk of 1 hr. from Lauffen along the Weissenbach leads to the spot. When the massive sluice-gates are closed, the water accumulates till it forms a small lake covered with floating timber. About once a week the gates are opened, of which notice is given at Ischl, and the rush of the water bearing its load of timber down the glen is a curious sight, well worth a visit.

The somewhat long list of excursions from Ischl here given by no means exhausts the roll, but those who spend a season here may be left to complete it for themselves. The mountaineer will naturally turn his eyes to the Dachstein peaks, which are further noticed in Rte. E.

ROUTE B.

LINZ TO ISCHL.

	Austrian miles	Eng. miles
Wels (by rly.) .	3½	16¼
Lambach . .	2¼	10½
Gmunden . .	3¼	15¼
Langbath (by steamer)	1¾	8¼
Ischl . . .	2	9½
	12¾	59¾

A large proportion of the Austrian visitors to Ischl, and nearly all of those from Bohemia and Prussia, approach it by way of Linz and the railway thence to Gmunden. This, the earliest of German railways, commenced in 1821, was designed and long used for horse-traction alone, the object being economy in the transport of salt from the salt-mines to the Danube. The portion between Linz and Lambach was made use of in the construction of the main line of rly. from Vienna to Salzburg and Munich—called in Austria Elizabeth-Westbahn. The remaining portion, from Lambach to Gmunden, is worked as a branch line by the same company. Not much has been gained by the substitution of steam for horse power, as the two trains that run daily from Linz to Gmunden employ 4 hrs. in accomplishing 42 m. The height of the railway bridge at Linz is 815 ft. above the sea, and the rise from thence to Gmunden 723 ft.; but the ascent is chiefly between Lambach and Gmunden. As far as the former place the rly. is carried over a plain country, at some distance from the l. bank of the Traun. There is little to attract attention save the distant outline of the Salzkammergut Alps, becoming gradually clearer and loftier as we approach them. The Traunstein and the Höllgebirg, as the nearest to the eye, assume an importance disproportioned to their real height.

Lambach (Inn: Rössel; there is also a new well-looking inn at the rly. station) is a small town with many substantial stone houses, famous for a Benedictine abbey, dedicated to the Holy Trinity, wherein the number three is repeated in every conceivable manner, even to the use of marbles of three colours for its pavement and decoration. The Ager, bearing the surplus waters of the Mond See and Attersee, joins the Traun close to the town. The branch rly., as well as the road to Gmunden, are carried along the rt. bank of the Traun. The *Traunfall*, or Fall of the Traun, nearly ½ way between this and Gmunden, is an object of attraction to many travellers. Those who take their tickets only from Linz to Lambach, and hire a carriage on arriving at the rly. station, may find time to visit the fall and go on to Gmunden in time for the steamer; charge for a carriage 5 or 6 flor. Others go by rly. to the Roitham station, and thence walk in ¼ hr. to the fall, returning to the same station for the next train, or walking on to Gmunden, nearly 9 m. distant.

The fall is caused by a broken ridge of conglomerate, running nearly all the way across the channel of the Traun, over and through which the stream breaks, falling through a vertical height of 44 ft. It is a cataract rather than a waterfall, resembling in some respects, but on a much smaller scale, the falls of the Rhine at Schaffhausen. The circumstance that the ridge of rock does not completely cross the Traun was turned to account in the 16th century in order to form a canal alongside of the river, by which the salt barges from Gmunden could pass to the Danube. For a fee of 70 kr. the miller will shut the canal gates, and thus send a large additional flow of water into the main channel, much improving the effect of the fall.

The scenery rapidly improves in interest as the traveller approaches *Gmunden* (Inns: Goldenes Schiff, by the lake; Goldener Hirsch; Goldner Brunnen; Sonne; all good and reasonable; besides several other second-class houses), a small town, beautifully situated at the N. end of the *Gmunden See* or *Traunsee*, the most stately of all the beautiful lakes of this district. It will remind the traveller of the middle part of the Lake of Lucerne, but the peaks of the Todtes Gebirg form a grander back-ground than the Mythen. Baths of brine, derived from the salt-mines, and a large hydropathic establishment, lead many visitors to reside here in summer, at little more than half the expense of a stay at Ischl. The finest views near the town are from the Tuschenschanze, the Calvarienberg, and from the Maxhügel. The first is especially to be recommended. The massive sluice gates, erected at the point where the Traun issues from the lake, deserve a visit. By their means the height of the lake is kept nearly uniform at all seasons.

The object which must here especially attract the attention of the traveller is the *Traunstein* (5,538'), rising above the E. shore, in almost vertical precipices, to a height of 4,200 ft. above the lake, about 4 m. S. of Gmunden. Though the absolute height be not very considerable, the extreme boldness of its form makes this one of the most remarkable mountains immediately overlooking any lake in the Alps; for though it does not approach the proportions of the summits enclosing the Bay of Uri, or the Lecco branch of the Lake of Como, or the neighbouring Lake of Hallstadt, its face is more precipitous, and it contrasts in a more remarkable way with the gentler slopes around. With the single exception of the Oetscher, near Maria-Zell (§ 54, Rte. F), this is the northernmost considerable summit among the outliers of the Alps. The outline of the mountain, as seen from Gmunden, offers a curious resemblance to the profile of Louis XVI.

Much nearer at hand is the castle of Ort, once belonging to Counts of that name, standing on an island SSW. of Gmunden, connected by a long bridge with the mainland. Though not remarkable in point of architecture, it is an extremely picturesque object. More interesting to the lover of art, is a visit to *Altmünster*, the most ancient village on the lake, about 2 m. from Gmunden. The church, containing several curious monuments, deserves examination.

Of excursions from Gmunden, that most recommended, is the tour of the Traunstein. A boat is taken to the foot of the mountain, and the traveller ascends thence by the Leinastiege, following a circuitous track to the *Laudachsee*, on the N E. side of the mountain. Thence the way back to Gmunden is by the Himmelreichwiese, enjoying at intervals very beautiful views of the lake. Active walkers, who do not desire to return to Gmunden, may go from the Laudachsee to Eisenau and Ebensee, but should have the aid of a local guide. The ascent of the Traunstein is easy, and practicable for ladies. The view is charming, though not equal as a panorama to that from the Schafberg (Rte. A).

The road along the W. shore of the lake has been only recently completed at a great expense, from the necessity for extensive blasting of rock. On that side, a succession of short promontories extend from the shore into the lake, and tend to diversify the scenery. On the first of these, after leaving Gmunden, is the village of *Ort*, opposite to the island mentioned above. Then the shore recedes to Altmünster, and projects again to a point whereon stands the castle of *Ebenzweier*. Near Fichtau is another point, and then follows a fourth, which, as seen from Gmunden, seems to close the head of the lake. Here stands the ancient convent of *Traunkirchen*, suppressed in the 16th century. Joseph v. Hammer has preserved, in a ballad, the traditional tale of a youth from the opposite village of Eisenau, whose illicit passion for a nun in this convent led him to emulate the exploit of Leander, and eventually to meet the same fate,

Traunkirchen enjoys the finest position on the lake shore, and the views from the surrounding heights, especially the *Sonnensteinspitz* (2,937'), are of extreme beauty.

Most travellers prefer to traverse the lake by steamer. Two of these ply regularly between Gmunden and Ebensee or Langbath, taking 1 hr. for the trip. The introduction of steam was here an important improvement, as the lake is exposed to sudden and violent storms, and boat navigation is attended by real risk in unsettled weather. The voyage is full of charm, and the interest attaching to the scenery is heightened when we regard this as the reservoir into which are gathered the streams from the other lakes that have made this region so attractive. Save the Mond See and Attersee, all join their waters here.

Ebensee, on the rt. bank of the Traun, is the chief place at the S. end of the lake. On the opposite side of the stream, close to the point where it falls into the lake, stands the village of

Langbath (Inn: Post), on the highroad to Ischl, and on that account, most passengers land here rather than at Ebensee. A large building, used for boiling the brine, of which a great portion is conveyed by pipes from Hallstadt and Ischl, is sometimes visited by tourists. Vast quantities of wood are carried down from the surrounding valleys; the smaller trunks and branches to serve as fire-wood, and the large stems to be formed into rafts that ultimately reach the Danube.

Langbath stands at the E. base of the range of the Höllgebirg (further noticed in Rte. C), and an interesting excursion may be made from hence to the *Kranabitsattel*. The shortest way is through the glen opening W. of the village. In 1 hr. the traveller reaches the Krehralp, and thence attains the summit in 3 hrs. But the active walker will make a detour by the lakes that lie in the Langbath glen above the Krehralp. In ½ hr. from thence he reaches the *Vordere Langbathsee*. The path along the S. shore is very rough, and if the fisherman be at hand, it saves time and labour to take boat to the upper end of the little lake, about 1½ m. in length. Another ½ hr. suffices to reach the second lake— *Hintere Langbathsee*—smaller, but more picturesque than the first. The *Kranabitsattel* is a limestone plateau, similar in character to the upper part of the Untersberg (§ 45, Rte. B). Its highest point is the *Feuerkogl* (5,175'). The view from this point has been lauded as amongst the very finest in this region.

Two other excursions, sometimes made from Ebensee or Langbath, deserve to be noticed here. The remarkable peak of the *Erlakogl* (5,051'), also called *Spitzlstein*, seen above the E. shore of the Traunsee, NE. of Ebensee, appears utterly inaccessible, but may be reached in 3½ hrs. by a steep, but not dangerous, path, and rewards the traveller by one of the finest views of the lake.

Between the Erlakogl and the Traunstein, two rather considerable torrents descend to the lake. One of these, called Röthelbach, from the red colour of the rocks above it, is seen to issue from a narrow cleft in the rocks. Mounting by a very steep goat-path, fit only for persons with steady heads, the traveller may reach a cavern called *Röthelbachhöhle*, in the interior of which is a little subterranean lake, spanned by a dome of rock. Those who go provided with Bengal lights enjoy a singular and striking scene, with a remarkakable contrast, when they, on issuing from the cavern, regain the view over the Traunsee and the surrounding mountains. To reach the path leading to the cavern the easiest way is to take a boat to a point on the shore near that where the Röthelbach falls into the Traunsee. This may be done as well from Gmunden as from Ebensee.

For the way from Ebensee to the Offensee, see § 47, Rte. D.

The road from Langbath to Ischl is traversed by omnibus twice a day, taking 2 hrs., but the scenery is so pleasing, that the traveller will prefer an agreeable walk of 3 hrs., unless he be hurried, in which case he may hire a carriage, with 1 horse for 3 fl. 15 kr., or with 2 horses for 5 fl. 25 kr., besides a trinkgeld of 50 kr. to the driver.

For Ischl and its neighbourhood, see Rte. A.

ROUTE C.

SALZBURG TO GMUNDEN, OR LAMBACH, BY THE MOND SEE AND ATTERSEE.

The traveller who has already approached the Salzkammergut by the high road from Salzburg, described in Rte. A., may take another longer but very interesting course, by which the Mond See, Attersee, and Traun See, will all be visited on the way from Salzburg to Ischl. It is also easy to unite this route with the former by taking the road from St. Gilgen to Mondsee, or by descending from the Schafberg to the latter village, and then following one or other of the courses described below; thus seeing everything of most interest in the exterior region before penetrating the interior recesses of Hallstadt and Aussee.

A sort of country diligence runs once a week from Salzburg, but at other times it is necessary to hire a carriage from that city to

Mondsee (Inns: Löwe; Krone), a village at the W. end of the lake. There is a bathing establishment on the shore, about 1 m. distant. The *Mond See* (1,626′) which takes its name from its somewhat irregular crescent form, is no unworthy rival of the neighbouring Wolfgang lake, from which it is separated by the range of the Schafberg. Though there is no single feature so remarkable as the Falkenstein, the mountains here rise more boldly, and the general effect is sterner and wilder.

An ancient Benedictine monastery at Mondsee, suppressed in 1791, was granted by Napoleon in 1810 to the Bavarian Marshal Wrede, in whose family it still remains.

Many pleasing excursions may be made from the village. The lover of lakes may follow the road leading NW. to *Zell*, and visit the pretty *Zeller See*, stretching northward nearly 4 m., between wooded heights, not to be confounded with its namesake in the Pinzgau, described in the last section. The stream from the Zeller See flows southward to Mondsee. Besides the Schafberg, noticed in Rte. A, several other adjoining summits offer views very beautiful, though less panoramic. The *Schober* (4,366′), which forms the western extremity of the Schafberg range, and lies between the Mond See and Fuschelsee, is easily reached in 2½ hrs., and well rewards the traveller's pains. On its western slope are the picturesque ruins of *Wartenfels*, which may be taken on the way going or returning. To NE. is the *Kulmspitze* (3,562′), commanding one of the best views of the Mond See and Attersee. Herr Hinterhuber, the apothecary of Mondsee, is an excellent botanist, and will doubtless aid with his advice a stranger investigating the flora of this neighbourhood.

No traveller visiting the lake will fail to take a boat excursion. If his object be to ascend the Schafberg, or to follow the road by the Krötensee to St. Gilgen (Rte. A), he will land at *Schärfling* where that road turns southward from the lake shore. If his course be to the Attersee, he will leave the boat at the E. end of the Mond See, and follow a track on the N. side of the See-Ache, which leads him in less than 2 m. to the larger lake, very near to *Unterach*, where there is a tolerably good country inn.

The *Attersee*, also called Kammersee, is the largest of the Austrian alpine lakes, being fully 12 m. long and in most places at least 3 m. wide. Its level is 37 ft. below that of the Mondsee, and its greatest depth is said to be no less than 1,597 ft. When seen by a stranger approaching this district from the N. (the nearest rly. station is *Vöcklabruck*, about 6 m. from the N. end of the lake) the effect is very fine, as the scenery constantly improves as the boat approaches the S. end. To one going by water in the opposite direction it is far less interesting. The shortest way, however, for a traveller wishing to reach Lambach is to hire a boat which will convey him from Unterach to *Schörfling* (not to be confounded with Schärfling, on the Mond See), and go thence by road to the *Vöcklabruck* station on the rly. from Salzburg to Linz, 9 m. from Lambach. Another way is to engage a country vehicle at Schörfling, which will carry him to the Traunfall (Rte. B), a distance of about 10 m. Near Schörfling, at the N. end of the lake, he will observe the castle of *Kammer* rising out of the water. It belongs to the Carinthian family of Khevenhüller.

If Gmunden be the traveller's aim, he should cross the lake diagonally from Unterach to Steinbach, on the E. shore, and thence follow a track through the hills, first E. and then NE., through the upper part of the *Aurachthal*, by which he may reach Gmunden in about 5 hrs. The traveller who would see something of the singular limestone plateaux, so characteristic of this region of the Alps, may be tempted to take, in his way from the Attersee to the Gmunden See, the range of the *Höll-Gebirg*, which extends almost continuously between those lakes. A considerable part of the range is so strictly preserved that no access is allowed to strangers not provided with a special permit. In any case, a good guide is indispensable, and such may probably be found at *Weissenbach*, a small village at the SE. corner of the lake, taking its name from a stream which is called Aeussere Weissenbach, to distinguish from the other stream descending towards the Traun, between the Ziemitz and Höllgebirg, which bears the name Mittlere Weissenbach. There is a very fair country inn at Weissenbach, which may serve as head-quarters for a traveller exploring the neighbourhood. The road hence to Ischl, following the two Weissenbachs, and the path leading by the Schwarzensee and Wirer's Wasserfall are noticed in Rte. A, Excursion 3.

The highest point of the Höllgebirg is called *Höllkogl* (5,754'), but the view is said not to equal that from the *Feuerkogl* (5,175'), the highest part of the so-called Kranabitsattel. [Kranabit, in the dialect of the Austrian Alps, means juniper, and the word recurs not unfrequently in local names.] In making excursions over this and similar high plateaux, ample allowance should be made for the time lost owing to the difficulty of the ground, which is intersected by rifts and wide fissures, and sometimes made more difficult by masses of creeping pine covering the surface. From the plateau of the Höllgebirg, the traveller may turn northward and reach Gmunden through the upper part of the Aurachthal, or descend to Langbath by the glen noticed in Rte. B.

Route D.

ISCHL TO STEINACH, IN THE ENNSTHAL, BY AUSSEE.

	Austrian miles	Eng. miles
Aussee	3¼	16¼
Mitterndorf	2	9½
Steinach	2	9½
	7½	35¼

Post-road. In starting from Ischl it is necessary to give 2 hrs.' previous notice to the postmaster. The traveller bound for the Styrian Alps, or for Bruck an der Mur. on the railway between Vienna and Trieste, will follow the road, here described. A portion of the way will be familiar to those who have made the excursions from Ischl described in Rte. A.

For about 3 m. the road from Ischl keeps to the W. side of the valley, but at *Lauffen* (Inn: Zum Weissen Rössel) it crosses to the rt. bank of the Traun. That stream here forms a cataract over a mass of broken rocks, but the salt-barges descend the rapid by a lateral channel. At the opening of the glen of the Ober Weissenbach, near Lauffen, the path to the Chorinsky klause turns off to W. Above Lauffen the valley of the Traun widens considerably, and in 3 m. more the road reaches *Goisern* (1,764'), a thriving village with two decent inns. The Protestants, who are numerous in all parts of the Salzkammergut, have here a church or meeting-house, whither English visitors sometimes resort on Sundays from Ischl. Above Goisern the valley of the Traun assumes a more alpine character; to the rt. rises the range of the Ramsauer Gebirge from which the Steinbach torrent descends to join the Traun near the village of *Ramsau*; on the l. is the Sandling, and beyond it the more imposing ridge of the Sarstein. In the centre the deep hollow filled by the Hallstadt Lake is defined by the steep mountains that inclose it on three sides. Before long the road to Aussee separates from that to Steg on the l. of Hallstadt, and after passing *St. Agatha*, begins to ascend the very steep though not lofty ridge of the *Pötschen Joch* (3,354'). This is a mere depression in the range which includes the Sandling and the Sarstein. The views on both sides of this pass are very interesting. Extra horses are taken for the ascent, which requires 1½ hr. in a carriage, or rather less on foot. The descent on the E. side is rapid, and the road soon reaches

Aussee. (Inns: Post, good but dear; Hackl, old-fashioned; Sonne, indifferent; beim Stüger; Blaue Traube.) A rather large village, 2,159 ft. above the sea, owing its existence chiefly to the neighbouring salt-works, but is much increased of late by the affluence of summer visitors drawn hither by the beauty of the neighbouring scenery, and in some measure by the salt brine baths which are administered on the same system as at Ischl, along with the whey-cure, &c. The position is not equal to that of Alt-Aussee or some other places in this district, but the immediate neighbourhood of several lakes, and the varied and striking forms of the mountains around, assure to the visitor a long list of varied and interesting walks, the more remarkable of which are noticed in Rte. A, Excursion 4. The finest panoramic view in the neighbourhood is undoubtedly that from the *Sarstein* (6,558'), lying between the mountain basin of Aussee and the Lake of Hallstadt. The ascent is sometimes, though rarely, made from Hallstadt, crossing the lake in a boat and mounting to the Sarstein Alp. The ascent from the Aussee side is steep and very rough, but in that way the effect of the grandest portion of the view is increased by the charm of surprise.

The Klamm, a spot especially interesting to geologists, is noticed in the next §.

The high mountains lying S. of Aussee, forming the E. portion of the great group of the Dachstein, are often distinguished by German writers as the Kammer Gebirge; the northernmost summit of this range is the *Hochkoppen*

(5,911′), which commands a view less panoramic, but quite as beautiful, as that from the Sarstein.

The way from Aussee to the valley of the Enns lies through a comparatively deep trough forming the natural division between the Dachstein Alps and the neighbouring group of the Todtes Gebirg. Near the village to SE. is a detached hill, the Radling, which partly fills the hollow space between the two groups. The post-road passes on the N. side of this eminence; while the *Oedenseer Traun*, descending from the clefts of the Kammer Gebirge, and especially from a tarn called *Oedensee*, flows along its SW. side, and joins the Ausseer Traun a mile below the village. The pedestrian will find it a more pleasing though rather longer course to follow the Oedenseer Traun, rejoining the post-road at Kainisch. Without any great ascent the road crosses from the basin of Aussee to that of Mitterndorf, which is scarcely inferior in point of scenery, and which is orographically important because the drainage is carried from it in three different directions. In the centre of this basin, which may better be described as an undulating plateau, is

Mitterndorf (2,638′), a village with a very fair inn at the Post. Above the village to N. and NE. rises a portion of the Todtes Gebirg; SW. lie the summits of the Kammer Gebirge; while the most picturesque and singular object in view is the *Grimming* (7,700′), an isolated mountain forming the eastern extremity of the main range of the Dachstein Alps, but separated from it by the deep defile of Stein. Through this opening the greater part of the drainage of the basin is carried southward to the Enns through the torrent called Salza, but not to be confounded with the more important streams of that name, one of which drains the Pinzgau, while the other joins the Enns at Reifling. A small tributary of the Traun flows from the W. side of the same plateau to join the stream from the Oedensee, and on the N. side of the Grimming most of the mountain torrents are united in the Grimmingbach, which runs ESE. to the Enns above Steinach. [The pedestrian who does not object to lengthen his walk by a few miles may well follow the rough track, just passable, but scarcely safe, for light chars, through the defile of *Stein*. This is so narrow that it is necessary to ascend to a considerable height above the torrent before commencing the steep descent towards St. Martin. The traveller here gains a fine view of the opposite range of the Styrian Alps, in which the Knallstein (8,511′) nearly due S., and the Wildstelle (8,998′) to SW., are the most prominent summits. He reaches the high road at St. Martin, about 6½ m. above Steinach.] Though not frequented by tourists, Mitterndorf must be a desirable stopping-place for a mountaineer, and especially for a geologist. The ascent of the Grimming is said to be best effected from *Klachau*, about halfway to Steinach. From its position it must command a remarkably fine view. Beds of lignite have been worked near the village.

For about 4 m. from Mitterndorf the post-road runs nearly at a level due E. until, on approaching the base of the Grimming, it turns to SE., close under the rugged face of the mountain. After passing the hamlet of Klachau (2,581′), it begins to descend along the rt. bank of the Grimmingbach, which has cut a deep cleft in the rocks. On the opposite side of the glen the Wallerbach forms a fine waterfall. Leaving to the rt. the castle of Neuhaus, opposite Irdning (§ 53, Rte. D), we join the high road from Radstadt, and soon after reach

Steinach (Inn: Post, tolerable country inn), in the valley of the Enns, farther noticed in the next section.

Route E.

ISCHL TO GOLLING, BY GOSAU. ASCENT OF THE DACHSTEIN.

A tolerable carriage road. About 40 miles.

The Gosau lakes, lying in an alpine glen that penetrates deeply into the mass of the Dachstein Gebirge, are constantly visited from Ischl, but may very conveniently be taken in the way from that place to Golling, in the valley of the Salza. The route here described is convenient for a traveller wishing to combine a visit to the Salzkammergut with a tour in Tyrol. Having gone from Salzburg to Ischl by the ordinary road, he may return westward by this way to Golling, thence reach Berchtesgaden by either of the routes noticed in § 45, Rte. F, and continue his journey by Saalfelden or Lofer. The road from Ischl to Gosau is much frequented, and in good condition. That from Gosau to Abtenau is rough, and it is advisable to walk part of the way. As every traveller halts at Gosau at least long enough to visit the lower lake, it is better to take the carriage from Ischl only to that place. A light carriage from Gosau to Abtenau costs about 5 fl., besides trinkgeld. The pedestrian should prefer the path from Gosau to Abtenau mentioned below, especially when the weather permits him to enjoy the noble view from the Zwieselalp. To the mountaineer Gosau is especially attractive as the most favorable point for ascending the Dachstein and the other surrounding peaks, and on this account the description of that high group is given here, although, as mentioned hereafter, it may be approached nearly as well from Hallstadt.

The road from Ischl to Gosau as far as the Gosaumühle on the l. of Hallstadt, a distance of rather more than 11 m., is noticed in Rte. A, Excursion 5. At the mill the road turns sharply to the l., and begins to ascend along the torrent descending from the Gosauthal. In 2 hrs. steady walking, or about 1½ hr. in a carriage, the traveller reaches the first of the numerous groups of houses that make up the village of Gosau. The *Gosauthal* consists of three portions very distinct from each other. The lower part, from the L. of Hallstadt to the village, is a narrow defile between the Ramsauer Gebirge to N.— whose highest summit, the Kahlenberg (6,066′), is often reached from this side (Rte. A, Excursion 10)—and the *Plassen* (6,403′), the chief of the northern outliers of the Dachstein group. This latter summit, which commands an admirable view, may very well be taken on the way from Gosau to Hallstadt. The middle portion of the Gosauthal is an open and comparatively populous valley, extending in all nearly 6 m. to the lower lake. Along the road leading to the lake are several small hamlets, besides many scattered solitary farmhouses. This tract rises from about 2,500 ft. at the lower end to rather more than 3,000 at the Vorder-See. The upper part of the Gosauthal is a wild alpine glen, running into the heart of the Dachstein group, and containing the upper lake, more than 4,000 ft. above the sea.

It is a matter of regret that there should not be more comfortable and convenient accommodation in the Gosau valley, one of the spots most attractive to the mountaineer in the Eastern Alps. At the lower hamlet—called for distinction *Vorder Gosau* (2,456′)—are two small inns. The first on the way from Gosaumühle (zum Dachstein) is uninviting in appearance; farther on (beim Brandwirth) are rough but clean quarters. Both are inconveniently far from the lakes and the higher mountains. Those who would give a day or two to explore this grand glen usually go to the farther hamlet—*Hinter Gosau*—where rough but clean rooms (often full in summer) and pretty good fare are found 'beim Hintern Schmied.' The host and his brother, Urstöger and Schnitzhofer, are the best guides for mountain excursions. Here, as at Hallstadt, the usual pay is 2 fl. per day, excepting for the Dachstein or Thorstein, for which 5 fl. is the ordinary pay. For a guide to Abtenau over the Zwieselalm, 1½ fl. is sufficient pay; and for a Tragsessel with 4 bearers, 7 fl. In half an hour from Hinter Gosau, or about 1½ hr. from Vorder Gosau, an active walker will reach the *Vorder Gosausee*. The char-road stops at the last houses, and visitors from Ischl accomplish the remainder of the way on foot, or in a *tragsessel*, by an easy path which mounts amidst scattered pines. The beautiful little lake, 3,051 ft. above the sea, about 1 m. long, and little more than ¼ m. wide, offers a view of unexpected grandeur. The Thorstein (9,677') forms the central point in the picture. The ridge connected with that peak, walling in the glen on the W. side, terminates in the remarkable summit of the *Donnerkogl* (6,731'), whose cleft vertex seen from the upper end of the Enns valley is there known as Bischofsmütze. On the opposite side the glen is inclosed by another ridge diverging from the main mass, the higher summits of which project through the surrounding mantle of glacier. The best point of view at the lower end of the lake is a covered shed where visitors lunch on the provisions brought with them, no supplies being available on the spot. A boat is usually to be found wherein the traveller may be ferried to the upper end, but in default of this conveyance there is a pleasant path along the W. side under the shade of old pines. There is a clean herdsman's hut near the upper end of the lake, where a mountaineer may pass the night tolerably well. Above the first lake the path mounts rather steeply through the narrow glen, and in 1 hr. reaches the

Hinter Gosausee (4,078'), little more than half the size of the lower lake, but offering a scene still grander and wilder.

Here the traveller feels himself face to face with the stern spirits that rule the upper ice-world. Even those who may not attempt any arduous excursions should not fail to reach this spot, quite easy of access, the expedition to and fro not requiring more than 5 hrs. from the inn at Hinter Gosau.

Ascent of the Dachstein. By way of introduction to the expedition which more than any other tempts the mountaineer in this part of the Alps, it seems desirable to give a slight description of the mass of mountains that, from the name of its highest summit, is called the Dachstein Group, or collectively the Dachstein Alps.

In all essential respects this bears a striking resemblance to the less extensive and less lofty range of the Steinerne Meer and Uebergossene Alp described in the last §. This also may best be described as an extensive plateau, well defined along the southern edge, and sloping thence irregularly towards the N. In the Dachstein Alps the summits are sharper, and project to a greater height above the general level, and the slope towards the N. is more rapid. For these reasons, and because a considerable part of the surface is covered with glacier, the plateau character is less evident, though in some places it is fully developed, and most of the strange phenomena that strike the traveller who crosses the Steinerne Meer may also be observed here. Although the southern slope of the range, extending nearly 30 m. from the Donnerkogl to the Grimming, is everywhere extremely steep, it is not very uniform in direction. The principal summits, indeed, lie nearly in a line drawn from W. to E., but about the centre some considerable masses project towards the valley of the Enns.

The irregularities of the surface, and even the more important prominences on the N. side, seem to be due to erosion rather than to the original form of the mass. It is therefore scarcely correct to speak of ridges projecting on that side from the central range; and if the term be used for convenience it must be understood to imply the mere fact that a series of eminences does exist, and affects certain directions.

Starting from the highest summit of the group — the *Dachstein* (9,845′) — which is at the same time the cornerstone of Austria proper, Styria and Salzburg, the main ridge extends WNW. Next to the Dachstein are the *Mitterspitz* (9,541′), and the *Thorstein* (9,677′), two extremely sharp obelisks of rock rising out of the glacier that covers the upper part of the ridge. Beyond the Thorstein the crest marking the boundary between the territory of Salzburg and the Salzkammergut is formed by the *Schneebergwand* (about 9,020′), and *Reissgang*. Thenceforward the range forming the W. boundary of the Gosauthal turns about due N., and it is connected with the Ramsauer Gebirge by the low ridge over which runs the road to Abtenau described below. Its chief summits are the *Hochgrumet, Graswandkogl* (7,909′), *Hochkogl* (7,050′), *Mannl* (7,112′), and the *Donnerkogl* (6,731′).

In the opposite direction to the Thorstein the E. portion of the main range diverges from the Dachstein. The chief summits going from W. to E. are the two peaks called *Diendeln* (or Fensterl), the *Koppenkarstein* (9,282′), the *Scheuchenspitz* (8,724′), and the *Koppeneck* (8,076′). The Koppenkarstein sends out a massive buttress to the S., inclosing on the E. side the *Todten Knecht* Glacier, the only considerable ice-stream descending towards the Enns. Between the Dachstein and Koppeneck the ridge marks the boundary between Salzkammergut and Styria, but thenceforward the boundary turns northward, towards the defile of the Traun and the ridge of the Sarstein, while the main ridge runs eastward to the *Stoderzinken* (6,926′), beyond which, after sinking towards the defile of Stein, it rises again to the bold summit of the Grimming (7,700′).

Of the ridges diverging northward from the main range, the most important is that connecting the *Nieder Dachstein* (9,645′) with the *Hoch Kreuz* (9,071′), and extending to the *Schöberl* (7.927′), and *Ochsenkogl* (7,142′). This divides the Hallstadt Glacier, or *Karls-Eisfeld*, from the Gosau Glacier and its tributaries. A serrated ridge projecting north-westward from the Hoch Kreuz to the *Hosswand* (8,193′) divides a small eastern branch from the larger central stream of the Gosau Glacier, which receives also a western tributary from a hollow on the N. side of the Thorstein.

The Hallstadt glacier is bounded on the E. side by a series of high summits diverging from the Diendeln, of which the most prominent points are the *Hoher Gjaidstein* (9,081′), and *Nieder Gjaidstein* (7,735′). These and the other ridges, if they deserve that name, extend farther, towards and beyond the Lake of Hallstadt, terminating towards the N. abruptly in the bold headlands that are the highest points seen from the shores of that lake. The chief of these are the *Krippenstein* (6,961′), the *Däuml* (6,561′), *Hierlatz* (6,426′), and *Zwölferkogl* (6,502′).

From the preceding description a correct general notion of the mass of the Dachstein Alps may perhaps be formed. A steep slope facing the S., with few recesses or projecting masses, and only one glacier of moderate size, contrasts with a NW. slope falling rapidly towards the head of the Glen of Gosau with a large central glacier and a small lateral glacier on either side, and with a much more extensive N. and NE. slope, partly occupied by a very extensive ice-field, but chiefly by an intricate and irregular mass of limestone rock, in some places lifted into sharp peaks and ridges, but usually presenting the aspect of a rifted plateau, from 4,500 to 7,500 ft. above the sea. This is seemingly almost bare of vegetation, yet it affords nourishment to numerous herds of cattle that are supported in summer from the vegetation that lurks in green hollows which do not strike the traveller's eye till he reaches the verge. Experience enables the mountaineer to find his way even in an entirely new district, because the key to local topography is found in the course of torrents, and their relation to the surrounding valleys. But on these limestone plateaux of the Eastern Alps this guidance altogether fails. The drainage, even from considerable glaciers, is carried away through subterranean channels. The extreme irregularity of the surface makes it impossible to follow a direct course, and the stranger is constantly liable to be stopped short by impassable clefts. Hence a guide is absolutely necessary for any considerable excursions in this group, and can scarcely be dispensed with even in shorter and easier ascents. The Dachstein group is perhaps more frequently visited from Hallstadt than from Gosau, and several of the less difficult expeditions are better made from the former place. Josef Seerainer, who lives at the Grüner Baum, is now considered the best of the Hallstadt guides. There are several others, of whom Ignaz Stocker, Zenner. Johann Schupfer, and Aloys Kirschleger have been recommended. Two fl. daily is the usual pay, but for the longer expeditions the guides expect from 7 to 9 fl. For the ascent of the Dachstein a traveller is expected to take two guides. Of the easier excursions from Hallstadt, the ascent of the Hierlatz, or, better still, of the Zwölferkogl, is recommended to those who would combine a view over the lake with a near survey of some of the higher summits.

A visit to the *Karls-Eisfeld* is also deservedly one of the favourite expeditions from Hallstadt. The easiest way is by the waterfall of the Waldbach Strub (Rte. A, Excursion 5), and thence by the source of the *Waldbach* (Waldbachursprung), where that torrent, which carries off the melting of the glacier, comes forth to-day after a long course underground. This is reached in 2 hrs. from Hallstadt, or equally soon by another way, passing the Rudolphsthurm and the Salzberg. Two hours more will take the traveller to the Wies-alp (5,482'), where explorers sometimes pass the night. More convenient quarters are found nearly 1 hr. higher up at the *Ochsenwiesalp* (6,023'). The shorter way is very steep; at one place the trunk of a tree leaning against the rock, and rudely notched, serves as a ladder. It is easy to ascend, but rather awkward for the descent. A hut built here by Seeauer of Hallstadt is very convenient for ascents undertaken from that place. It is at no great distance from, and but little lower than, the end of the glacier; but the ground is so rough that it is a long day's work to reach the glacier and return to Hallstadt. It is a better plan to sleep at the Ochsenwiesalp, and if bound for Gosau the traveller may reach it by a very grand pass. The way lies by the *Schladmingerloch*, and Grünberg. Passing close to the Hosswand, the traveller descends to the Hinter-See in 7 hrs. from the Ochsenwies alp.

No better proof need be given of the steepness of the higher peaks of this group than the fact that of the 10 highest summits only four have yet been climbed, and that of these only one—the Hoher Gjaidstein—is reached without difficult rock-climbing. Access to the summit of the Dachstein from the Hallstadt side has of late years been rendered comparatively easy by the aid of two or three steps cut in the rock, and by iron rings driven into the stone, to which ropes have been attached, for most of which assistance mountaineers are indebted to Prof. Simony, of Vienna. Among other writings connected with this district he has published a very interesting account of a winter visit to the Karls-Eisfeld, reprinted in the new edition of Schaubach's Deutschen Alpen. The way to the summit from Gosau is shorter than from Hallstadt. In 5 hrs. the peak may be attained from the sennhütte above the Hinter-See, and the traveller may return to the inn at Hinter Gosau in the evening; but the passage along the rocky arête is far more difficult than the Hallstadt route, and would be dangerous to any but a steady and practised mountaineer. To such the best course would be to ascend by one route and descend by the other. The summit of the Dachstein is visible from Ischl, Alt-Aussee, the Gosau lakes, and many points in the Ennsthal, and these, as well as many other more distant valleys, are included in the view from the peak. The mountain panorama is very extensive and interesting, extending to the Schneeberg near Vienna, and the Terglou, on one side, and on the other to the Zillerthal Alps. When seen from the valleys on the N. side, the mountain is liable to be confounded with its nearer rivals, and on this account it was at one time supposed that the names Thorstein and Dachstein belonged to one and the same peak. When the distinction between them was fully established, it remained for a while uncertain whether the Thorstein were not the higher of the two, though it has since been proved to be lower by 168 feet.

The ascent of the *Thorstein* is said to be decidedly more difficult than that of its higher rival, owing to its extreme steepness, and to masses of ice that sometimes (probably only in early summer) cling to the ledges by which the peak is attained. The first ascent was accomplished by Jacob Buchsteiner, a hunter from Schladming. 'None of the present guides know the way to the summit.' [E. M.] The other high peaks of this group were all considered inaccessible until 1862, when the Hoch Kreuz was climbed by a hunter from Hallstadt. His companion, daunted by the difficulty of the enterprise, turned back before reaching the summit.

The writer is disposed to recommend to the attention of mountaineers visiting this district the passes already known, as well as those which remain to be discovered, across the higher parts of the range. These, if undertaken in favourable weather, do not appear to be difficult, and they must offer much of novelty and interest. There is one track, known to the local guides and chamois hunters, leading from Obertraun, at the head of the L. of Hallstadt, to Schladming, and crossing the main range E. of the Koppenkahrstein. A more direct pass may be made from Hallstadt to the same place by mounting to the head of the Karls-Eisfeld, and descending on the opposite side by the Todten Knecht glacier. The summit of this pass is reached in making the ascent of the Dachstein from Schladming. Another fine pass from Hallstadt to Ramsau, over the ridge between the Dachstein and the Diendeln, is said to have been made by a Styrian hunter. Having crossed by one or other of these passes, the traveller may return by a comparatively easy way from Schladming to Gosau by the *Reissgang Pass*, some way W. of the Thorstein. Besides the higher and comparatively laborious pass leading from the Ochsenwiesalp to the upper end of the Gosauthal above the lower lake, there is a much easier way, requiring only from 6 to 7 hrs. from Hallstadt, by which the traveller may descend very near to the inn at Hinter Gosau.

There are two ways by which the traveller may go from Gosau to Abtenau. The char-road crosses the lowest part of the ridge, connecting the Dachstein with the Ramsauer Gebirge by the *Pass Gschütt* (3,247'). The distance is reckoned 4 hrs. on foot, or 3¼ hrs. in a country carriage, the road being very rough, and the ascent in some places steep. Though not comparable to the path by the Zwieselberg, it affords some charming views, which indeed can scarcely be missed in this beautiful district. On commencing the descent, the range of the Tännen Gebirge, not seen hitherto by the traveller who has wandered through the Salzkammergut valleys, presents a fine appearance to the W. (see Rte. G). The descent leads through the narrow glen of the Russbach, an affluent of the Lammer. The rocks here are rich in fossils, and fine specimens may sometimes be bought for a trifle from the men who are employed in mending the road. From the junction of the Russbach with the Lammer, the road mounts a hill to reach Abtenau.

The path from the Gosau valley to the last-named village, most recommended to pedestrians, turns westward from the track to the Vorder-See a little above Hinter-Gosau. A finger-post with the direction 'Annaberg und Radstadt' points the way. The path winding up the mountain can scarcely be missed, and in little more than two hrs. from the Hinter-Schmied's inn the traveller reaches the châlets of the *Zwieselalm*, Zwieselalp, or Zwieselberg. Here mountain fare and rough night-quarters are found in case of need. In a few minutes' walk from the Hütten the rounded grassy summit of the Alp is attained, and here at a very moderate height (about 5,000') a panoramic view of extraordinary extent is obtained. Besides the nearer groups of the Dachstein, Tännen-Gebirge, and Berchtesgaden Alps, nearly the entire range of the Hohe Tauern from the Ankogl to the Gross-Venediger, is seen to the utmost advantage. If bound for Werfen or St. Johann, the pedestrian should go nearly straight down the mountain side to some sennhütten, whence a beaten track leads to Annaberg (Rte. F); but if his object be to reach Abtenau, he will do better to follow a slightly marked path leading NW., by which he may fall into the road from the Gschütt Pass, close to the junction of the Russbach and Lammer.

Abtenau (Inns: Post, good, not cheap; Ochs) is finely situated, 2,336 ft. above the sea, at a considerable height above the Lammer, which in its course to join the Salza near Golling has excavated singular hollows in the rock, similar in character to those of the Oefen above Golling (§45, Rte. E). To S. and SW. rises the range of the Tännen-Gebirge (Rte. G), and in its N. face is a large cavern, called *Frauenloch*, sometimes visited from Abtenau.

On Mondays, Thursdays, and Saturdays, a country carriage carrying letters goes from Abtenau to Golling, starting at 12½—charge 1 fl.—and returns on the following days from Golling at 8 A.M. The charge for an einspänniger wagen is 3½ fl. and trinkgeld, the distance being counted as 4 Stunden. The road bears to the l., or WSW., towards the base of the Tännen-Gebirge, and then descends a long and steep hill to a bridge over the Lammer. Thenceforward the road runs along the rt. bank of that stream, traversing *Scheffau*, a village with an interesting 14th century church, containing paintings attributed to Wohlgemuth, the master of Albrecht Dürer. The road enters the valley of the Salza between the village of Golling and the Oefen, and the traveller may well visit that curious locality (§ 45, Rte. E) before turning northward along the high-road to Golling.

The pedestrian who has no occasion to visit Abtenau may follow a more interesting and rather shorter way to Golling by keeping along the rt. bank of the Lammer from its junction with the Russbach. The ravine of the Lammer is curious, and of itself deserves a visit; but that of the *Aubach*, a stream which joins it from the N. about 1½ hr. below the Russbach, is still more singular. The cleft cut by that torrent is so narrow that a block fallen from the mountain on either side has been stuck fast between the opposite walls, and forms a natural bridge 200 ft. above the level of the torrent. A little farther on the Aubach forms a fine waterfall—the *Bichlfall*—which may be taken in the way without any great detour. There is a track from Abtenau, reaching the Lammer a little above its junction with the Aubach, by which the traveller may visit the Bichlfall, and traverse the ravine of the Aubach without losing more than 1 hr. on his way to Golling.

The limits of this work do not permit any details respecting the geology of the Dachstein-Alps and the valley of Gosau. Our distinguished countrymen Sedgwick and Murchison were among the first to throw light upon the intricate relations of the secondary strata here exhibited on so vast a scale. The fuller development of these relations, and the establishment of the order of succession of the different beds, has been accomplished by the patient researches of several Austrian geologists, to whose memoirs, especially those of Reuss and von Hauer, the geologist visiting this district should not fail to refer.

Route F.

ABTENAU TO RADSTADT, OR WERFEN, BY THE FRITZTHAL.

5½ Stunden, or about 13 m., to St. Martin—7½ St., or 17¼ m., thence to Werfen—4 St., or 9¼ m., from St. Martin to Radstadt.

Although the Salza is the only stream that carries the waters from the N. side of the central range of the Eastern Alps to the plain of S. Germany, within the wide space between the opening of the Zillerthal and the point where the Enns reaches the plain near Steyer, the mountain masses forming the outward border to the N. are separated by deep depressions, which are usually made use of for the passage of a carriage road. In § 45 we noticed the depression between Zell-am-See and Saalfelden, which all but allows the waters of the Salza to flow through the channel now traversed by the Saale, dividing the Lofer Alps from the Berchtesgaden group. In the same way it was seen that the depression between the Dachstein group and the Todtes-Gebirge allows the passage of the road from Aussee to Steinach described in this section, Rte. D. A similar depression to that last mentioned divides the Tännen-Gebirge from the Dachstein. The traveller starting from Golling by the road described in the last Rte., ascends the valley of the Lammer along the N. side of the range of the Tännen-Gebirge till, on reaching Abtenau, he sees to SW. the wide opening between that and the Dachstein group through which the Lammer flows from the S. A rather rough, and little used road leads from Abtenau across the low ridge closing the head of the valley of the Lammer, and joins the post-road referred to in § 45, Rte. E, which connects Werfen with Radstadt in the valley of the Enns.

That road passes along the N. side of a rather singular triangular mass bounded on the W. by the Pongau, and on the S. by the road from St. Johann to Radstadt. This latter southern road marks the prolongation of the great trough of which the western portion is occupied by the Salza, and the eastern half by the Enns, and the isolated mass whose highest summit is the Grundeck (5,949'), is in fact the eastern prolongation of the zone of transition rocks which were seen to occupy a wide space between the crystalline rocks and the secondary limestones near Kitzbühel, and a more contracted area in the neighbouring district of Dienten. The road here described enables a traveller going S. from Abtenau either to reach Radstadt on the Enns, or Werfen in the valley of the Salza, by a short and agreeable route.

As the eastern extremity of the Tännen-Gebirge lies due S. of Abtenau, the road goes at first due E. to avoid the spurs from that mass, and then, after crossing the Lammer, ascends to *Annaberg*, where, at the village inn, a traveller arriving from Gosau by the Zwieselalm may usually find conveyance. Hence for two stunden the road continues to ascend gently amid a wooded hilly tract, with the noble summits of the surrounding mountains visible at intervals. The little village of St. Martin stands close to the summit of the ridge, and commands a very fine view to the S. The transverse valley through which the road from Werfen goes to Radstadt is called *Fritzthal*. Its torrent flows westward, and enters the Salza opposite Bischofshofen. The traveller bound for Radstadt takes the easternmost of two roads leading from St. Martin into the Fritzthal, and almost immediately after reaching the level of the glen crosses the stream and follows the road across the low ridge that separates it from the head of the Enns valley. He reaches that river at Altenmarkt, and 2 m. farther arrives at the post-station of *Radstadt*, further noticed in § 47, Rte. A.

Instead of crossing the ridge on the S. side of the Fritzthal the pedestrian bound for Schladming may follow the Fritz to its head at *Filzmoos* (3,355'), and then descend to join the high-road at the defile of Mandling. This scarcely involves any loss of time, and takes the traveller very near to the base of the higher peaks of the Dachstein range.

In going from St. Martin to Werfen, the traveller takes the l. hand, or westernmost, of the two roads from the former village, and soon falls into the high-road about 1½ m. E. of *Hüttau*, the post-station between Werfen and Radstadt, 2½ Austrian m. from the former, and 2 Aust. m. from the latter place. The early existence of this line of road through the Noric Alps is proved by a Roman milestone preserved in the church. Below Hüttau the valley of the Fritz becomes a narrow glen, and no distant object meets the eye for about 7 m., when the road emerges on the slope above the valley of the Salza, having left the Fritz to make its way to that river through a ravine on the l. hand. The view over the Salza valley, with a portion of the mass of the Uebergossene Alp in the background is extremely striking. The road descends diagonally to Dorf Werfen, then crosses the stream, joins the high-road from Lend, and soon reaches the market-town of *Werfen*. See § 45, Rte. E

Route G.

ABTENAU TO WERFEN BY THE TÄNNEN-GEBIRGE.

The mountain mass called Tännen- (or Tennen-) Gebirge, lying east of the Salza and separated from the Hagen-Gebirge (in the Berchtesgaden group) by the defile of Lueg, may be roughly described as a parallelogram 12 m. in length from W. to E. by 8 m. in breadth from N. to S. The mass is pretty nearly enclosed by the streams of the Fritz, Salza, and Lammer, leaving only the low ridge mentioned in the last Rte., whereon stands the village of St. Martin, which partially connects it with the Dachstein group, but its geological structure shows that its original connection was with the Berchtesgaden group through the Hagen-Gebirge.

The mass consists of a high central barren plateau approaching to 7,000 ft. above the sea-level, and subsiding both on the E. and W. side to hollow basins fully 1,000 ft. lower in level, which afford pasture in the summer. From the E. and W. extremities the higher summits rise like watch-towers to guard the central mass. The most prominent of these are the *Bleikogl* (7,905') over Abtenau, the *Wieselsteinkopf* (7,537') at the NW. end towards Golling, and the *Raucheck* (7,967') at the SW. corner overlooking Werfen. The latter is the highest summit of the entire mass. The general character of the group is the same as that of the Todtes-Gebirg, described in the next §, but this is more difficult of access, and is very rarely visited, except by the few herdsmen who resort hither in summer. Guides are hard to find, and the only one recommended is Joseph Schorn of Abtenau. With fine weather a mountaineer should be able to cross the plateau from Abtenau to Werfen in one long day. He may best descend into the defile of the Salza by the *Steinerne Stiege*, a very steep path by which the W. face of the plateau is made accessible. It would scarcely be practicable to take the summit of the Raucheck on the way from Abtenau to Werfen, except by passing a night at one of the huts on the mountain. Excepting the Vorder Pitschenbergalm, these are poor and comfortless. See a paper by Baron Guido v. Sommaruga in the *Jahrbuch* of the Austrian Alpine Club, vol. ii.

SECTION 47.

ENNS DISTRICT.

In the introduction to this chapter the valley of the Enns, running eastward from Radstadt to Hieflau, and then northward towards the Danube, was indicated as the natural limit of the mountain region here collectively termed the Salzburg Alps. Having in the last section described the Dachstein group, and the contiguous valleys and lakes of the Salzkammergut, there still remains a larger but less lofty district, separated from the last by the valley of the Traun and the road from Aussee to Steinach, and bounded on the S. and E. by the valley of the Enns. This district is subdivided into two nearly equal portions by the valley of the Steier and the depression over which runs the road from Lietzen to Windischgersten (Rte. B). To the W. of that boundary lies the very extensive mountain range collectively known as the Todtes-Gebirg, from its exhibiting on the largest scale those bare limestone plateaux to which that characteristic name is given in Styria. This extends from the Hohe Schrot (5,691') near Ischl due eastward to the Kleiner Priel (6,995'). Southward from the main range a region of high plateaux is continued for many miles to the neighbourhood of Aussee and Mitterndorf. From near the latter village a southern division of the same mass spreads eastward to the *Warscheneck*, (8,112'). The highest summit of the group is the Grosser Priel (8,238'), lying near its N E. corner. The utmost length of the group from the Hohe Schrot to the Warscheneck exceeds 30 English m. On the E. side of the Steier the mountains are broken up into small masses separated by relatively deep valleys, and their height diminishes as we travel eastward. The most prominent of these separate mountain masses are the Hoher Bürgas (7,351') N. of Admont; the Buchstein (7,269'); the Sengsen-Gebirge, culminating in the Hoher Nock (6,459'); and the Gross-Alpkogl (4,953).

Although the superior attractions of the adjoining Salzkammergut district make this comparatively neglected, even by German travellers, it is by no means deficient in natural charms, and the ranges and plateaux of the Todtes-Gebirg in particular offer scenery of the wildest and strangest character, and supply abundant occupation to the geologist and the naturalist. The traveller may gain an idea of the scenery of the

Todtes-Gebirg by excursions from Alt-Aussee or Dürrenbach, but to enjoy it thoroughly he should spend at least one night at one of the châlets on the plateau. Admont is, however, the most desirable head-quarters in this district, and two or three weeks may well be devoted to the surrounding country.

The name *Enns District* seems that most suitable to this region, although the upper extremity of the Ennsthal falls beyond its limits. The limestone peaks S. of Admont are described in § 53.

Route A.

ENNS, ON THE DANUBE, TO RADSTADT, BY THE ENNSTHAL.

	Austrian miles	Eng. miles
Steyer	3	14
Losenstein	3	14
Weyer	3	14
Altenmarkt	3	14
Hieflau	2¼	12
Admont	3¼	16½
Lietzen	2½	11¾
Steinach	2	9¼
Gröbming	2½	11¾
Schladming	2	9½
Radstadt	3	14
	30¼	142

Railway from St. Valentin, on the Westbahn railway, to Weyer. Post-road thence to Radstadt.

The portion of the Rudolfsbahn railway between Weyer and Hieflau is nearly complete, and the line will ultimately pass very near to Lietzen. (See § 53, Rte. E.) A carriage with two horses (zweispann) can be hired from Hieflau to Admont for 7 fl., and thence to Lietzen for 6 fl. From Admont to Aussee the charge is 18 fl. These rates include toll and *trinkgeld* to the driver. The route may be shortened by about 12 m. by taking the road from Altenmarkt to Admont by St. Gallen, but this avoids the finest scenery of the Enns valley.

Enns is a very ancient town, about 13 m. from Linz, on the rly. from that place to Vienna. Its authentic history commences when it became a Roman military station, already famous for the manufacture of arms from the iron mines of the adjoining district. Several hundred years later the Bavarians erected here a fortress, to resist the advances of the Huns. The walls of the town are reported to have been built with the money paid to Duke Leopold of Austria for the ransom of Richard Cœur de Lion. The church and town-hall contain several objects of antiquity, but the neighbouring church of *Lorch* is still more interesting in this respect.

Those who approach the valley of the Enns from the north side will naturally go by railway from the St. Valentin station, near Linz, by the branch line to Steyer, where the Enns issues from the last low undulations of the Alps into the plain. The first portion of the Rudolfsbahn railway, for which see § 53, Rte. E, is now open as far as Weyer.

Steyer (Inns: Goldener Löwe; Ochs; Schiff). This is a rather considerable town, with thriving manufactures of iron and steel, from the excellent material afforded by the mines in the upper valley of the Enns. The town stands at the junction of that stream with the *Steyer*, from which it takes its name. It is rather singular that the adjoining province of Styria (in German, *Steyermark*) has taken its name from that river, although the entire course of the stream, as well as the town, lie in Upper Austria. The 15th century church, with good coloured glass, and a tower commanding a fine view, deserves a visit. The castle of Count Lamberg, on a height overlooking the town, and a large building on a hill over the suburb of Steyerdorf, must command still more extensive views. About 9 m. W. of Steyer is *Hall*, now rather frequented on account of its mineral spring, containing iodine in rather large proportion. A diligence runs daily from Hall to Steyer, and thence to the St. Peter station on the rly. to Vienna.

The road leading from Steyer through the valley of the Enns to Hieflau, and thence to Eisenerz, is commonly called the *Eisenstrasse*, or Iron Road, as by this way the larger part of the produce of the important mines of Upper Styria is conveyed to the plain. Almost every village is in some way engaged in the same business. Smelting works, foundries, forges, or some accessory establishments meet the eye in every inhabited spot. Following the road along the rt. bank of the Enns, about 9 m. S. of Steyer, the valley begins to assume a somewhat alpine character. Approaching the base of the *Hoch Buchberg* (4,169'), it turns to ESE., and preserves that direction for the ensuing 18 m. The post-station is at

Losenstein, a village in which the exclusive occupation of the people seems to be making nails. About 3 m. farther is *Arzberg*, opposite to which, on the l. bank, is *Reich Raming*, with extensive iron-works, the largest of which belongs to the Austrian government. A considerable deposit of lignite, of the age of the lias, extends eastward from this place to Waidhofen.

The geologist may study the succession of the secondary deposits, from the lias to the neocomian formation, by following the course of a little stream called the Pechgraben. The intelligent officers who direct the imperial foundry will give useful local indications to the geological inquirer. On a height above Arzberg is an inn, standing near a tree, said to be the largest oak in Austria. It commands a fine view of the valley. The post-station of

Weyer stands at the opening of a lateral glen, through which a good road is carried ENE. to Waidhofen (§ 54, Rte. F), 15 m. distant.

Weyer is the head-quarters of the iron district of Upper Austria, and the residence of the officials who superintend the numerous government establishments. The scenery of the Ennsthal, which has been constantly increasing in interest, attains the climax of picturesque beauty throughout the stage of 14 m. between this and the next post-station. The road turns nearly due S., and ascends along the rt. bank of the river, between the range of the *Gross Alpkogl* (4.953') to the W., and that of the *Voralpe* (5,632') to the E. The little stream of the Frenz, descending from the latter range, and that of the Laussabach on the opposite side of the valley, mark the boundary between Austria and Styria. Almost immediately after entering the latter province, the valley again turns eastward, and in the angle formed by the stream stands the little town of

Altenmarkt (Inn: Adler, good), not to be confounded with the place bearing the same name mentioned below, at the head of the Ennsthal. For the road hence to Windischgarsten, see Rte. E. The shortest way from Altenmarkt to Admont and the Upper Ennsthal is by *St. Gallen* and *Buchau*, a distance of 17 or 18 m., but the lover of nature will much prefer the road following the Enns through one of the finest defiles of the Eastern Alps. It has been made passable for carriages only of late years.

Between Altenmarkt and Admont the Enns flows in a very circuitous course round the S. and E. sides of an isolated mountain, whose highest summit is the Gross Buchstein (7,269'). This forms a nearly triangular mass, separated from the range of the Bürgas and Mureck to NW. by the glen of Buchau, through which runs the road from Altenmarkt by St. Gallen to Admont, forming the third side of the triangle.

The road along the Enns, for the first time since leaving Steyer, crosses to the l. bank at *Reifling*. Close to this village the Styrian Salza pours its stream into the Enns from the picturesque valley leading to Maria-Zell (§ 54, Rte. B). A massive grating, nearly 2,200 ft. in length, crosses the stream just above the junction, to arrest the vast quantities of floating timber felled in the Salza valley for the consumption of the forges and foundries of this district.

From Reifling the road follows for some time the l. bank, but returns to the opposite side of the stream at *Hieflau* (1,538′), a small village with two or three inns, of which the best is Steuber's, opposite the post. The position of this village is of some orographic importance. It lies at the eastern extremity of the great trough, extending hence to the head of the Pinzgau, which marks the northern limit of the crystalline rocks constituting the central chain of the Tyrol and Styrian Alps; which trough, as was seen in the introduction to this chapter, is nearly continuous with that long reach of the Inn valley which in the same way marks the northern boundary of the crystalline zone in Western Tyrol. It is true that between this place and Admont the calcareous secondary strata trespass across the boundary, and attain a great height on the S. side of the valley; but they scarcely extend beyond the precipitous masses which the traveller sees from the road, and this local exception can scarcely be held destructive of the significance of the facts above pointed out. The ranges enclosing the parallel valley of the Styrian Salza, may perhaps be regarded as indications of the action of the same forces.

A light carriage, carrying the post, plies daily between Hieflau and Lietzen —fare, 2 fl. 60 kr.

Up to this point, the traveller has followed the so-called Iron Road, keeping a general direction towards SE. Here the Enns takes the nearly due westerly direction, which it preserves to the head of the valley, henceforward called Ober-Ennsthal, while the Iron Road follows the glen of the Erzbach to the famous mining town of Eisenerz, and leads thence to Leoben (§ 53, Rte. H).

The *Defile of Gesäuse*, through which the Enns forces its way from Admont to Hieflau, is justly famed as one of the grandest and most picturesque scenes in the Alps. Some others may be named where the absolute height of the mountains above the level of the stream is greater, but few where the precipices rise more boldly to so great a height. The bottom of the defile is densely wooded wherever there is space for trees to take root, and above the forest rise on either side shattered towers and pinnacles of secondary limestone, attaining a vertical elevation of nearly a mile above the roadway. This was long a mere foot-path carried from one side to the other on ricketty wooden bridges, and it is only since 1850 that the well-constructed new carriage-road has been open to travellers. The Enns, which here becomes a furious torrent, especially in early summer, descends through the defile in a succession of falls and rapids a vertical height of 460 ft., hurrying down pine trunks, felled on the surrounding mountains, that are arrested at Hieflau by a strong grating similar to that at the mouth of the Salza opposite Reifling.

Less than 1 m. beyond Hieflau the road passes a group of houses called Ennsbrand, and fairly enters the defile. On the northern side the river is walled in by the almost continuous mass of the Buchstein, merely broken by a few narrow clefts, through one of which, about half way from Hieflau to Admont, it is possible to attain the summit of the mountain. On the S. side the still higher range overlooking the defile culminates in the Hoch-Thor (7,478′), a summit which has preserved the reputation of inaccessibility, though it was reached by M. Schleicher in 1854. On the l. hand is seen a ravine called *Hartellsgraben* running deep into the range of the Hoch-Thor.

About 3 m. farther a much deeper opening on the same side admits the sunshine into the heart of the defile. The considerable glen of the Johnsbach here pours its torrent into the Enns. Its source is on the S. side of the Hoch-Thor, and after flowing westward for some miles, and passing the picturesque village of *Johnsbach* (2,400'), it turns northward, dividing the Hoch - Thor from its western rival, the Reichenstein. For a further notice of the Johnsbacherthal see § 53, Rte. I.

Nearly opposite the confluence of the Johnsbach is the cleft on the opposite side of the Enns, by which it is possible to make the ascent of the Buchstein, passing by the Gstatterboden. After traversing the grand and beautiful scenery of the defile for about 10 m., the road finally issues into the open valley of the Enns near a solitary inn called Heinlbauer (1,997'). For several miles the valley is open and level, and the waters of the Enns, held back by the narrow Gesäuse passage, move with a slow current, and have partly overflowed the adjoining land, converting it into marsh where it has not been reclaimed by the labour of the monks. Midway in the valley, about ½ m. S. of the Enns, and 2,005 ft. above the sea, stands

Admont (Inns: Post, kept by Dräxler; Buchbinder's; both afford very fair quarters). This place derives its name and its importance from the famous Benedictine monastery (ad Montes) founded by Archbishop Gebhard of Salzburg in the 11th century. Long celebrated for its wealth and luxury, and more honourably for the literary and scientific labours of some of its members, the community has in the present century experienced a period of comparative poverty, chiefly owing to the rapacity of Gallic invaders. As in some other Austrian monasteries, the monks have in some degree kept pace with the general movement of the age. They have been active agricultural improvers, and in the ecclesiastical college, or seminary, maintained here, instruction in agronomic science and other practical matters is given to the students, with a view to the future advantage of the poorer classes in remote districts where the monks have the cure of souls. The very extensive pile of buildings, dating from the 17th century, as well as the church, were almost completely destroyed by fire in 1865. The library, built only about 90 years ago, was rescued from the flames. It is a really fine hall, richly decorated, and containing a precious collection of manuscripts and incunabula (including an unique copy of Ottokar's Chronicle in rhymed verse), besides 20,000 vols. of printed books. The museum, including very complete collections of the local fauna and flora, was lost at the same time. The fish-ponds, with separate reservoirs for each species of fish, covered in and locked up, are among the curiosities of the place.

Admont affords excellent head-quarters to the naturalist and the mountaineer; and the lover of legendary lore may amuse himself in collecting tales of fairies and kobolds that still circulate among the peasantry. The village suffered severely in the great fire of 1865. The adjoining marshes are said to breed fever in autumn, and the frequency of cases of cretinism among the inhabitants seems to indicate the presence of malaria: at other seasons there can be nothing to apprehend on this score. Those who do not undertake longer excursions should visit the Mariakulm church for the sake of the noble view which it commands.

The excursion most to be recommended to the mountaineer is the ascent of the *Grosse Buchstein* (7,269'), the highest point in the mass, lying between the road from Admont to St. Gallen and Altenmarkt and the angle formed by the Enns between those places. As already mentioned, it is accessible by a very steep path through a ravine on the N. side of the defile of Gesäuse. Those who take that way usually sleep at a very fair mountain inn at the Gstatterboden, and make the ascent next morning. An easier way is to follow the road to St. Gallen which crosses the *Weng-Pass*, and descends on the NE. side through the glen of Buchau. From a little mountain roadside inn (Am Eisenzieher), about 1 hr. above St. Gallen, the ascent direct to the summit may be made in 3½ hrs. J. Reidegger is a good local guide, and such is required by a traveller descending by an opposite side from that taken in the ascent. The view, besides including a very extensive alpine panorama, commands a long reach of the Ober-Ennsthal; but perhaps the most interesting object is the deep defile opening immediately under the ridge of the mountain, with the remarkable pinnacle of the Hoch Thor rising at the opposite side of the cleft. The Hoher Bürgas (7,351′) is also within reach of this place, but may more conveniently be visited from the Pyrhn Pass (Rte. B), or from Rosenau (Rte. E).

For an account of the fine peaks rising S. of the Enns the reader is referred to § 53.

The botanist will find ample occupation in this neighbourhood: among other rarities he may collect *Calla palustris* and *Andromeda polifolia* in the marshes below the village, where *Ledum palustre* also once grew, and on the surrounding mountains *Papaver alpinum*, *Draba stellata*, *Cherleria imbricata*, *Potentilla minima*, *Saxifraga Burseriana*, *Valeriana elongata* and *V. celtica*, *Achillea Clusiana*, *Cineraria alpestris*, *Saussurea pygmœa*, *Crepis hyoscridifolia*, *Pedicularis asplenifolia* and *P. rosea*, several rare *Salices*, *Malaxis monophylla*, *Epipogium Gmelini*, *Allium Victoriale*, *Festuca Schuchzeri*, and *Sesleria tenella*. *Cortusa Matthioli* has been also found in clefts of the limestone rocks in the defile of Gesaüse.

There are two roads from Admont to Lietzen, of which the post-road, best suited for carriages, is by the l. bank of the Enns. The fine views of the neighbouring mountains are better seen in approaching from the west, than when travelling in the opposite direction. To the l. of the *Mühlauthal*, which descends from the Hoher Bürgas, and pours its torrent into the Enns N. of the village, is seen the pilgrimage church of *Mariakulm* on an isolated hill; and behind it, farther W., the rounded ridge of the *Pleschberg* (5,611′). After crossing the river the road turns westward, and runs along the base of the Pleschberg to *Ardning*, where a torrent issues through a narrow gorge from the *Ardningthal*.

[The mountaineer may reach Ardning by a circuitous but very interesting walk by ascending the Mühlauthal from Admont, then crossing the ridge which connects the Pleschberg, with the Hoher Bürgas, and descending through the Ardningthal to rejoin the high road. If bound for the valley of the Steyer, he may cross a second ridge close to the SW. base of the Bürgas, and join the high road from Lietzen (Rte. B.). near the summit of the Pyrhn Pass.

The valley of the Enns widens out to a breadth of fully 2 m. as the road, after traversing *Reitthal*, passes opposite the junction of the Paltenbach with the Enns. Through the Paltenthal runs the road leading by Rottenmann and Mautern to Leoben and Bruck (§ 53 Rte. E). The traveller joins that important line of communication between Salzburg and Lower Styria at the post-station of

Lietzen (Inns: Stanzinger's, best; and several others), a thriving little market town, 2,168 ft. above the sea, where the stranger may procure good specimens of Styrian iron and steel manufacture. Another road, already in use in the Roman period, leads north eastward along the Pyrhnbach to the valley of the Steyer (Rte. B). An extensive peat moss near Lietzen occupies a portion of the space formerly filled by a shallow lake.

From hence to Gröbming the valley bends slightly to S. of W. This portion of the Ober Ennsthal is remarkable for numerous ancient castles which add much to the picturesqueness of the scenery. The road, keeping to the l. bank, passes at the little village of *Weissenbach* the opening of a wild glen through which one of the many streams bearing the same name issues into the main valley. The glen is completely enclosed at its head by the crags of the *Angerhöhe*, one of the summits forming the eastern extension of the Todten-

Gebirg. The traveller who carries a sketch book will scarcely fail to halt a while here, and will be tempted to do so again about 3 m. farther on, as he passes the ruined castle of *Wolkenstein*, cradle of the still extant family of that name. The building seems to be of one piece with the limestone rock on which it stands, which here assumes a red tint on long exposure to the weather.

Below the castle is the village of *Wörschach*, by the opening of another glen which contributes its torrent to the main stream. At Niederhofen the road passes by a little church reported to be the most ancient in Styria. On a rock above is a castle destroyed by lightning within the last 20 years. The owner has replaced it by a modern residence in the Swiss style, offering a curious contrast with the older building. A little way beyond the village of

Steinach (Inn: Post, small but tolerably good) the road to Ischl and Salzburg (§ 46, Rte. D), descends from the N. side of the Grimming to enter the Ennsthal. This is one of the most picturesque points in the Ennsthal, and the traveller who does not object to rather rough country quarters may well halt for two or three days at the little inn at Steinach. At the base of the singularly fantastic crags and pinnacles of the Grimming is the castle of *Trautenfels*, rendered habitable by its present owner. Opposite to it, on the S. side of the river, is the village of *Irdning* (2,197'), at the opening of the Donnersbacherthal, interesting to the botanist, by which the pedestrian may ascend the Hohenwarte or cross the pass leading to Oberwölz (§ 53, Rte. D).

The road to Radstadt runs along the S. base of the Grimming till, at St. Martin, the Salza torrent issues from the defile of Stein on the W. side of that mountain (see § 46, Rte. D). Opposite to this is the opening of the Walchernthal; and a little farther on the traveller looks into the more considerable glen of the *Sölkthal*, with the fine peaks of the Knallstein (8,511') in the background (§ 53, Rte. C).

Gröbming (Inn: Post, good) stands on a plateau above the N. bank of the Enns, 2,466 ft. above the sea, on the Gröbmingbach, a torrent descending from the Stoderzinken (6,926') which rises NE. of the village. The church deserves a passing visit; it contains some curious old pictures. The road for some miles keeps over a terrace or lower shelf of the Dachstein range, somewhat above the level of the valley; but at *Aich* it descends towards the Enns, and soon crosses to the rt. bank. After traversing two torrents descending from the Wildstelle, the traveller reaches *Haus*, a small village with a fair country inn, and about 4½ m. farther gains the post-station at

Schladming (2,402'), a finely situated village with a good inn (Post), which was utterly destroyed in 1529 in revenge for the successful revolt of the people of this district, and again half consumed by an accidental fire in 1814.

On the N. side of the valley, high above the green hills of the Ramsau, rise the bare grey precipices of the Dachstein, and invite the mountaineer to a nearer approach. Three passes leading to Hallstadt are noticed in § 46, Rte. E. A much easier, but long, day's walk to Gosau may be made by *Filzmoos*, and thence, by the Hochalm, over a pass called Scharwandalm, more commonly *Das Steigl* (6,737'). The best plan is to sleep at Filzmoos, visiting the Ramsau by the way. The first ascent of the Thorstein was made from this side by a chamois-hunter of Schladming, named Jacob Buchsteiner. The Schladmingthal, leading to the Hoch Golling, is described in § 53, Rte. B. Johann Bachler and Matt. Lechner are recommended as guides.

In approaching Schladming the traveller will note a change in the style of the buildings. Here, and throughout the upper end of the valley, the houses have the nearly flat roofs, formed of slit pine, kept in place by large blocks of stone, prevalent throughout the Tyrol

Alps; while through the lower valley the more picturesque high roofs, with metal ornaments, are nearly universal.

Two easy excursions, which may be combined in a single day's walk, are strongly recommended to the traveller who would fully enjoy the scenery of the upper valley of the Enns. The first is the ascent of the *Planaykopf*, due S. of Schladming, the summit of which is easily reached in 3 hrs. It commands a view reaching from the head of the Ennsthal to the W. to the neighbourhood of Admont in the E. Close at hand is the great range of the Dachstein, here called simply Stein, showing a succession of towers and precipices such as are not easily matched elsewhere in the Alps. In the opposite direction, beyond the Enns, is the great chain of the Tauern rising from the Gumpeneck and Knallstein to SE., to its highest summit—the beautiful pyramidal peak of the Glockner in the SW. Less expected by the traveller, who feels that he has reached but a very moderate height, are the summits of the Uebergossene Alp and the neighbouring mountains of the Berchtesgaden district, including the double peak of the Watzmann. But the greatest surprise to one who has viewed the Dachstein range from Schladming, where it seems to rise immediately above the first line of low green hill that borders the Enns, is the tract called the Ramsau, extending westward for 8 or 9 miles from the Planaykopf, with a breadth of from 3 to 4 miles. This is an irregular plateau extending along the base of the Dachstein, and terminating towards the S. in the green declivity already alluded to.

Those who may not ascend the Planaykopf should, in any case, make an excursion to the *Ramsau*, remarkable for its position at the very base of so great a line of mural precipices, as well as for the peculiar character of the population. These are nearly all Protestants, whose ancestors held their creed in secret until the toleration edict of Joseph II. allowed them to profess it. The story that they descend from refugees who escaped the Salzburg persecutions is denied on the spot. A wide breadth of tillage land, besides numerous flocks and herds, support a people whose very existence is unsuspected by those who pass along the main valley.

Resuming his journey by the highroad, the traveller, on leaving Schladming, returns to the l. bank of the Enns, and after passing opposite the opening of the *Preuneggthal*, which penetrates due S. into the Tauern range, in about 7 m. from the village reaches the so-called *Pass Mandling*, forming the limit between the territory of Salzburg and Styria. This is merely a narrowing of the Enns valley at a point where the *Mandling* torrent, marking the ancient frontier, descends from the N. to join the Enns. The Mandling is formed by the confluence of a stream arising from the melting of the snow in the crevices of the Dachstein, and therefore called the Kalte Mandling, with another from the NW., called by way of contrast Warme Mandling. The pedestrian who has made an excursion from Schladming to the Ramsau may rejoin the road to Radstadt by following the stream of the Kalte Mandling to the pass.

On his way hence to Radstadt the traveller will perceive, by the forms of the mountains on the S. side of the valley, that at this point the orographic boundary does not exactly coincide with the geological; and, in fact, a fringe of dolomitic mountains of moderate height here forms the northern margin of the Tauern range. Ascending slightly, the traveller reaches the curious little walled town of

Radstadt (Inn: Post, reports as to prices and accommodation are discordant), standing at the junction of three of the most important roads of the Eastern Alps. One of these leads by Hüttau (§ 46, Rte. F) to Werfen, and thence (§ 45, Rte. E) to Salzburg; another is that of the Radstadter Tauern (§ 52, Rte. E), forming the main line of com-

K 2

munication from the N. into Carinthia; while the third is that described in the present Rte., by which the traveller may either pursue the Enns to its junction with the Danube, or follow through the Paltenthal from Lietzen the road to Leoben and Bruck an der Mur. A fourth road leads westward by Wagrein to St. Johann-im-Pongau (§ 52, Rte. E), and affords the most direct way to the Pinzgau and to Bad Gastein. The town, standing 2,564 ft. above the sea, is enclosed by a high wall, with but two gates, one at the east—the other at the west side, was almost completely destroyed by fire in 1781, and again in 1865, so that its aspect does not correspond with its antiquity. The church tower and a gothic chapel in the cemetery are among the few remaining objects of interest.

It was preceded by another more ancient town, believed to occupy the site of the Roman Ani, which still stands about 3 m. distant on the S. bank of the Enns. This was long called Alt Radstadt, to distinguish it from the comparatively modern town, but is now known as *Altenmarkt*, not to be confounded, however, with the place mentioned above in the Unter Ennsthal.

The chief source of the Enns is a stream that flows northward from the Tauern Alps, through the glen of Flachau, to *Reitdorf,* about 5 m. W. of Radstadt, there turns eastward, receives the *Zauchbach* from a short glen S. of Altenmarkt, and at Radstadt is doubled in volume by the junction of the Tauernache from the Radstadter Tauern. (See § 52, Rte. E).

ROUTE B.

STEYER TO LIETZEN, BY THE PYRHN PASS.

	Austrian miles	Eng. miles
Leonstein	3½	16¼
Dürrenbach	2½	11¾
Spital-am-Pyrhn	3	14
Lietzen	2	9½
	11	51¾

(A post-road very little frequented.)

The valley of the *Steyer*, which joins the Enns at the town of Steyer mentioned in the last Rte., has been unduly neglected by strangers. As far as Dürrenbach, a road of high antiquity, probably in use in Roman times, runs along the stream, and there turning SW. to Windischgarsten, reaches the valley of the Enns across the low ridge of the Pyrhn Pass. Though not offering any such remarkable scenery as that of the defile of the Enns below Admont, this road is quite equal in attractions to the remaining portion of that described in the last Rte., and to a traveller approaching the upper valley of the Enns from the E. or NE., it offers a short cut of at least 32 Eng. miles. As mentioned below, the valley of the Steyer is also easily reached from Wels by a good post-road.

At Sirninghofen, about half-way on the road from Steyer to Hall (see last Rte.), the river Steyer, which has hitherto flowed nearly due E., is seen to issue from the south through the wooded hills that form the outer skirts of the northern Alps. The road here turns SSW. along the l. bank of the stream, and occasional glimpses of the grey peaks of the Todtes Gebirg add to the interest of the way. Although not rivalling the importance of those in the Enns valley, the iron works here show considerable activity. At *Steinbach,* where the valley is contracted between the rocks, pocket knives are produced in large quantities; and some miles farther on, at Mölln, the fabrication of Jews'-harps is the chief occupation of the inhabitants. The last-named village stands on the rt. bank of the Steyer, at the junction of the *Krumme Steyerling*, a torrent flowing from the E. side of the Sengsen Gebirge, mentioned below. By that way the pedestrian may reach Windischgarsten in as short a time as by the post-road; or by bearing eastward from the head of the glen, he may strike a country road leading to Altenmarkt in the Unter Ennsthal.

Opposite to Mölln, on the l. bank of the Steyer, is *Leonstein*, the post-station. The valley here bends to SW. until about 5 m. farther, by the church of Frauenstein, an opening to the northward, with scarcely any perceptible barrier, connects the valley of the Krems with that of the Steyer.

[By that way a post-road leads from the railway station of Wels by *Voitsdorf* (2½ Aust. m.), to *Kirchdorf* (2 Aust. m.), whence the post-station of Dürrenbach is 2¾ Aust. miles distant. This is the shortest way for a traveller approaching Styria from Munich and Salzburg, or from central Germany by Ratisbon. The distance from Wels to Lietzen is 12¼ Austrian, or 57½ English miles, and from the latter place he can reach Bruck an der Mur, either by Admont and Eisenerz, or by the rather shorter road over the Rottenmanner Tauern.]

Beyond the junction of the road from Wels, the valley of the Steyer ascends due S. for several miles through the Defile of Klaus. This does not rival in grandeur those of the Enns, the Salza, or the Saale, but the scenery is throughout very pleasing; numerous groups of houses, and two castles, occupy the more open spaces which recur at intervals as the road passes between the W. end of the range of the *Sengsen Gebirge*, and the less lofty limestone ridges, broken by several ravines and narrow glens, that enclose the valley on the l. side. At the S. end of the defile is the post-station of

Dürrenbach, also written Diernbach. S. of this village is the *Damberg* (4,966'), which divides the upper valley of the Steyer into two branches. The stream which preserves the name Steyer descends on the W. side of the Damberg from a deep recess in the range of the Todtes Gebirg, while the lesser branch, called *Teichel*, flows from SE. along the E. side of the former mountain. The road to the Ennsthal follows the l. bank of the Teichel through the latter branch of the valley. The church of St. Pancraz is conspicuous from a distance in the narrow glen through which the road ascends gently till, about 5 m. from Dürrenbach, it opens into a wide basin, surrounded by wooded slopes. Here stands, amid very charming scenery,

Windischgarsten, the chief place on the road, but not a post-station, with three fair inns—Fuxjäger, König von Sachsen, and Gemse. The landlord of the last is a good mountaineer. Here diverges a rough country road to Altenmarkt (Rte. E). The way to Lietzen lies due S., through a narrow opening between the Warscheneck (8,112'), which crowns the SE. extremity of the Todtes-Gebirg, and the Hoher Bürgas (7,351'). Following a chain of little lakes, the road, about 6 m. from Windischgarsten, reaches *Spital-am-Pyrhn*. In the 12th century a hospice was founded here for the use of pilgrims to the Holy Land, who then commonly followed this road through the Alps. At a later period converted into a college, this was suppressed in 1807; and the church, with some rather good pictures, has since been neglected. Nearly 3 m. beyond Spital is another fine church at *St. Leonhard*. Bearing to the rt., or SW., immediately under the steep rocks of the Hoher Bürgas, also written Pyrgas, the road ascends amid wild scenery to the *Pyrhn Pass* (3,162'), forming the frontier between Austria and Styria. On the way the traveller passes near to the *Schreibachfall*, a waterfall, which if the volume of water were greater would have attained to celebrity, as the total height is not less than 1,200 ft. An excursion to the Gleinkersee and the Pieslingursprung is much recommended. Soon after crossing the ridge the traveller gains an extremely fine view of the range of the Styrian Alps S. of the Ennsthal, and then descends rapidly to Lietzen, noticed in Rte. A.

As mentioned in Rte. A, the pedestrian bound for Admont may save several hrs.' walk by following a track from the Pyrhn pass, over the so-called *Bürgas Sattel*. From the summit of that pass the peak of the Hoher Bürgas is reached in less than 2 hours.

Route C.

DÜRRENBACH TO AUSSEE, BY THE TODTES-GEBIRG.

In the last Rte. it was observed that the principal branch of the Steyer issues near Dürrenbach from a glen that runs deeply into the mass of the Todtes Gebirg. By that way the traveller may conveniently visit a group of mountains hitherto neglected by tourists, in consequence of the superior attractions of the neighbouring range of the Dachstein, yet very interesting from the wildness and singularity of its scenery, which nearly resembles that of the Steinerne Meer, described in § 45, Rte. D. In approaching the Ischl district from Vienna, the way over the Todtes Gebirg, though rarely traversed, is very convenient for the mountaineer who does not shrink from a long day's walk over ground of the roughest description, and, irrespective of other attractions, the geologist and the naturalist will find ample occupation by the way. The upper valley of the Steyer may also be taken in the way from Dürrenbach to Lietzen; for, as mentioned below, the traveller may either return to the highroad at Windischgarsten, or follow a direct course to Weissenbach in the Ennsthal, only 2 m. W. of Lietzen.

Above its junction with the Teichel, near Dürrenbach (see last Rte.), the Steyer issues from a cleft between the *Kleiner Priel* (6,995′) and the Damberg. In the narrowest part of the defile, the stream forms a fine waterfall, 87 ft. in height, to see which to advantage, the traveller should turn aside from the track to reach a point below the fall. Before long the defile opens into an alpine basin, whose green pastures, dotted with substantial farm-houses, are encompassed by the rugged masses of the Todtes Gebirg. To the E. is a recess in the mountains, where stands the hamlet of

Vorder Stoder, whence a cart-track crosses the ridge of the Hocheck, and descends to Windischgarsten. From Vorder Stoder the mountaineer may reach the summit of the *Waschenegg* (8,112′), the highest summit of that branch of the Todtes Gebirg, which encircles on the E. & S. sides the head of the Steyer. By the main stream is *Hinter Stoder*, one of the most characteristic of the secluded villages of the Austrian Alps, where the naturalist who would give a few days to exploring the neighbouring mountains may best fix his headquarters. In attempting any expeditions amidst the upper plateaux he should be accompanied by a local guide, for mountaineering experience serves but little amid the strange irregularities of the surface, and in case of an accident, he might lie for weeks or months without aid from human hands.

The head of the valley, wherein the streams from the surrounding mountains unite to form the Steyer, is an irregular square space, measuring 5 or 6 m. each way. Hinter Stoder lies at the NE. angle of the square, of which the principal torrent, rising at the opposite angle, forms the diagonal. On every side this basin is enclosed by the Todtes Gebirg, whose outlines appear from below as mountain ridges, but are usually no more than the abrupt edges of the surrounding plateaux. Massive buttresses project into the quadrangular space forming the drainage basin of the Steyer, and separate the torrents which are often concealed by masses of limestone débris, and come to light only when they reach the level of the lower valley. Waterfalls of great height abound, but except after heavy rain the supply of water is insufficient.

The traveller, bound for the Ennsthal, should follow the main torrent till, after crossing a tributary that issues from a considerable lateral glen to the S., he ascends by the buttress lying in the fork between the two streams. By keeping a general course due S. across the summit of the ridge, he will finally reach a track that descends rapidly eastward to the Weissenbach, and will join the high-road of the Ennsthal at the village of that name, 2 m. W. of Lietzen. The least laborious way to Aussee is to follow the main stream of the Steyer, and ascend by a difficult path to the high plateau, whereon lie two tarns —the *Steyersee* and *Schwarzensee*—without visible outlet, but supposed to feed the Steyer through subterranean channels. A path descends SW. to Tauplitz, a hamlet about 4 m. from *Mitterndorf* (§ 46, Rte. D), whence Aussee is reached by the high-road.

The mountaineer, who would form a closer acquaintance with this singular district, should not hesitate to give two, or even three, days to an excursion amidst the upper region of the Todtes 'Gebirg, sleeping at night in the *Sennhütten*, which are far cleaner and more habitable than is usual in Switzerland, or on the S. side of the Alps. If he be moderately conversant with the language, the opportunities for personal intercourse with the fine population of these remote recesses of the Austrian Alps will add much to the interest of the expedition. It will be understood that the term plateau, as applied to the upper regions of these mountains, is not to be understood in a literal sense. The surface is not distributed into ridges and depressions, through which streams run in a definite course; but, on the other hand, nothing can less resemble a plain than the greater part of the extensive tract here in question. Its characteristic feature is the recurrence of deep caldron-shaped hollows (Germ. *Kessel*), often of great extent, sometimes forming lakes, but more frequently carpeted at the bottom by green pastures. Hither the herdsmen resort in summer, and were it not for the intimate local knowledge thus acquired, they could scarcely succeed in guiding a stranger amidst perplexities of ridge and hollow that might compete for intricacy with the Cretan labyrinth. At intervals project from above the general level the higher summits of the group; and the traveller should not omit to ascend at least one of these, in order to form a true notion of the general character of the surrounding tract, and to enjoy the noble views of the Dachstein group and the Tauern Alps, which are always obtained in clear weather. It is not easy to indicate any particular course as more interesting than another, but the mountaineer will probably prefer to include in his excursion the highest summit of the range —the *Grosser Priel* (8,238′), once erroneously supposed to be the highest of all the mountains between the Salza and the Enns. The most direct way is through a narrow glen, called *Polsterthal*, opening less than a mile above Hinter Stoder. At its head is an amphitheatre, or *cirque*, enclosed by precipitous mountains, of which the highest is the *Spitzmauer* (7,954′). Here, and farther on, the traveller passes many a waterfall, offspring of the Staubbach family, but quite unknown to fame. Bearing to the left from the Posterthal, through a second hollow, the mountaineer who started in the morning from Dürrenbach, will reach before evening the *Klinseralp*, where he will do well to fix his quarters for the night. Starting early next morning, and taking the shorter way by a snow slope, and up steep rocks, he may reach the summit of the Grosser Priel in 2 hrs.' steady walking. In descending he should keep nearly due W. to the Rothgeschirr, and thence bear SW. to the *Elmsee* and *Elmgrube*. On the way it is well worth while to diverge to the rt., and gain a point in the ridge of the Röllscharte, whence, standing on the summit of a range of tremendous precipices, he looks down on the little *Almsee* lying at the N. foot of the range (see Rte. D). The Elmgrube, a deep hollow surrounded by high crags, is reached in 6 hrs. from the summit of the Priel. From hence the traveller may descend in about 2½ hrs. to the little inn (beim Ladner) on the *Grundelsee* (§ 46, Rte. A, Excursion 4), passing two picturesque tarns—the Hintere, and Vordere, Lahngangsee. The active mountaineer will prefer to give a day or two more to exploring this singular region, and, if time allows it, may push on from the Elmgrube by a rough and intricate path, with many an intermediate ascent and descent, to the Henaralp or the Brunnwiesenalp, whence he may on the following day visit the Klamm. This latter spot was visited many years ago by two eminent English geologists, who have given the following description of the appearance of the rocks:—' The face of the Grossberg, a mountain of secondary limestone, which shuts out the valley of Klam from the Grundlsee, is singularly scooped out into grooves and furrows, which, wherever the surface is nearly vertical, are straight, semicircular, and deeply engraven; but, where the limestone sweeps down in a slope, they are wider and shallower, and increase in number, branching out from each main trunk like gigantic arms, with expanded and pendent fingers. No drawing or description can convey more than a faint idea of the extraordinary contortions and dislocations of the rocks which surround the little upland valley of Klam.'—*Murchison and Sedgwick.*

From hence the traveller may descend to the Töplitzsee, and thence reach Aussee by the Grundelsee. Instead of going from the Elmgrube to the Henaralp, it would have been a shorter course to steer for the Vordernbachalp. On the following day the traveller may take the Klamm on his way, and pass a third night at the châlets of the *Brunnwiesenalp*, which may be considered the pastoral centre of the plateau. The merry *Sennerinnen*, from the surrounding alps, are used to gather here towards evening, especially when it is known that strangers are arrived, a rare event, and keep up dance and song till a late hour. From hence the traveller may make a circuit by the Wildenseealp and the Klopf to Alt-Aussee. (See § 46, Rte. A, Excursion 4.)

The traveller who does not care to ascend the Grosser Priel, may much shorten the way from Hinter Stoder to the Elmgrube, going direct in one day from Dürrenbach, by a steep but not difficult track which passes the Schneethal. In that case he should not fail to ascend one or other of the summits that overlook the general level of the plateau. Of these, the *Röll* (7,218′) and the *Hochelm* (6,966′), are probably the most advisable. The latter is particularly easy of access, the former interesting, from its lying on the precipitous northern edge of the plateau.

One of the best centres for excursions in the Todtes Gebirg is the *Augstwiesenalp*. ' From thence the ascent of the *Schönberg* (6,502′) is strongly to be recommended. Travelling over the plateau in an easterly direction for 3 hrs.—passing the Wildenseealm and the Wildensee—the highest eastern peak is reached in 1 hr. more. The panorama is the finest in the Salzkammergut, except that of the Schafberg. The peak is also easily reached from Aussee by the Aussee-Rettenbachalm, by a path called the Nagel, or from Ischl by the Ischl-Rettenbachalm. Both paths meet at the Schwarzenbergalm.' —[E. M.]

It will be understood that these expeditions require a steady head, and some practice in rock climbing. Delicious fish are found in many of the tarns of these mountains, and the traveller who does not arrive too late may hope to improve his supper by a bargain which is usually easily concluded.

An extensive cavern in the Kleiner Priel, called *Kreidenlucke*, may interest amateurs. It seems to be unusually difficult of access.

Route D.

DÜRRENBACH TO ISCHL OR GMUNDEN.

Between the Steyer and the Traun a mountain district of considerable extent lies between the steep northern face of the Todtes-Gebirg and the low country bordering the Danube. As was observed among the outer ranges of the Bavarian Alps, the prevailing direction of the ridges and the minor glens is from E. to W.; but the greater part of the drainage of the district in question is carried northward through a glen which seems to have been cut transversely through these ridges. The stream, called *Albe*, or Alm, joins the Traun below Lambach.

Respecting the district here spoken of the editor has obtained but little information, although it certainly includes much picturesque scenery, and may well tempt a pedestrian wishing to go from the valley of the Steyer to Ischl or Gmunden.

At the N. end of the defile of the Steyer above Dürrenbach, mentioned in the last Rte., a track mounts westward, and passes along the northern base of the *Teufelsmauer*, which is the ridge connecting the Kleiner with the Grosser Priel. On reaching the northern base of the latter peak, the path passes two small tarns, and follows a stream, issuing from a cleft in the mountain, which flows NW. to its junction with the Albe.

Here the traveller has to decide whether his object be to reach the valley of the Traun between Ebensee and Ischl, or to go direct to Gmunden. The former is the more interesting walk. If he choose this, he must turn sharply to the l., and follow the Albe southward for about 1 m. to its source in the *Almsee*, a small lake about 1 m. long, lying very near the foot of a grand range of precipices forming the northern face of the Todtes-Gebirg. Accommodation for the night may probably be had at a solitary building called Seehaus at the N. end of the lake.

The way to Ischl is through a short lateral glen that opens WSW. at the lower end of the Almsee. Bearing due W. from the head of this glen the traveller will reach the *Offensee*, another small and picturesque mountain lake, sometimes visited by Ischl tourists. Above it to the S. rises the *Rinnerkogl* (6,254'), one of the Todtes-Gebirg range. A track practicable for light country carriages leads from the lake down the picturesque glen of the *Frauenweissbach* to its junction with the Traun, about 2½ m. above Langbath, or 7 m. from Ischl. If the traveller wish to reach Ebensee, he should follow a country road which turns to the rt. from the Frauenweissbach some way before its junction with the Traun. About 8 hrs., exclusive of halts, must be allowed for the walk from Dürrenbach to the Offensee, and 3 hrs. thence to Ebensee.

The traveller bound for Gmunden, who has reached the upper end of the Albe valley by the way already described, should on no account omit to visit the Almsee before he commences his walk down the stream. The upper part of the valley is called *Grünau*, from the principal village bearing that name, which is reached by a cart-road keeping chiefly to the rt. bank of the Albe. It is about 2½ hrs. walk from the Almsee to the village, which has an ancient church deserving a visit.

The easiest way from Grünau to Gmunden is to follow the road along the Albe to *Mühldorf*, and then turn westward by a hilly road leading past St. Conrad to Gmunden. The pedestrian will, however, do much better by taking in his way the picturesque little *Laudachsee*, mentioned in § 46, Rte. B, lying on the NE. side of the Traunstein. It possesses a remarkable echo. The walk thence to Gmunden commands charming views of the Traunsee.

At Mühldorf a road turns aside from the Albe valley, and leads NNE. by the very ancient village of Pettenbach to *Kremsmünster*, famous for one of the wealthiest and most stately of the Benedictine abbeys of Austria. It combines the magnificence of a royal palace with a library, museum, and observatory, that would do honour to any university. The observatory is on the top floor of a tower nearly 200 ft. in height.

The Offensee, mentioned above, may be reached without difficulty from the Wildenseealp, and so taken in a day's walk between Aussee and Ebensee. 'There exists also a way from the plateau of the Todtes Gebirg to the Almsee, but that is said to be dangerous.'—[E. M.]

The way lies a little S. of E. from Windischgarsten to *Tambach*, about 4 m. distant, and then begins the rather steep ascent of the Rosenleite leading to *Rosenau*. The ridge traversed by this road, dividing the waters of the Steyer from those of the Enns, extends northward from the *Hoher Bürgas* (7,351') to the *Wasserklotz*. From a country inn called *Eckel im Reith*, near the summit level, the ascent of the former peak may be effected. The descent towards the Enns lies through the *Laussathal*, whose torrent marks the boundary between Austria and Styria. It joins the Enns close to Altenmarkt (Rte. A).

The pedestrian bound for the lower part of the Enns valley may shorten the way by taking a track that turns northward from the upper part of the Laussathal, traverses the ridge E. of the Wasserklotz, and descends due N. along the *Reich-Ramingbach* to the village of that name on the Enns. Another way is to make the ascent of the *Gross-Alpkogl* (4,953'), the highest summit of the mass lying between the Reich-Raming glen and the Enns. Either course is interesting to the geologist. See Rte. A.

Route E.

WINDISCHGARSTEN TO ALTENMARKT IN THE LOWER ENNSTHAL.

This way is practicable for a light country carriage, and may be recommended to a stranger who wishes to see something of a very unfrequented district of Upper Austria.

BALL'S ALPINE GUIDES,
LATEST EDITIONS.

The Alpine Guide. By JOHN BALL, M.R.I.A. late President of the Alpine Club. Post 8vo. with Maps and other Illustrations, in Three Volumes, as follows:—

The Guide to the Eastern Alps, price 10s. 6d.

The Guide to the Western Alps, including Mont Blanc, Monte Rosa, Zermatt, &c. price 6s. 6d.

Guide to the Central Alps, including all the Oberland District, price 7s. 6d.

Introduction on Alpine Travelling in general, and on the Geology of the Alps, price 1s. Either of the Three Volumes or Parts of the *Alpine Guide* may be had with this INTRODUCTION prefixed, price 1s. extra.

To be had also, for the convenience of Travellers visiting particular Districts, in Ten Sections as follows, each complete in itself, with General and Special Maps:—

THE BERNESE OBERLAND, price 2s. 6d.

MONT BLANC and MONTE ROSA, price 2s. 6d.

DAUPHINÉ and PIEDMONT, from Nice to the Little St. Bernard, price 2s. 6d.

NORTH SWITZERLAND, including the Righi, Zurich, and Lucerne, price 2s. 6d.

The ST. GOTHARD PASS and the ITALIAN LAKES, price 2s. 6d.

EAST SWITZERLAND, including the Engadine and the Lombard Valleys, price 2s. 6d.

NORTH TYROL, the Bavarian and Salzburg Alps, price 2s. 6d.

CENTRAL TYROL, including the Gross Glockner, price 2s. 6d.

SOUTH TYROL and VENETIAN or DOLOMITE ALPS, price 2s. 6d.

The STYRIAN, CARNIC, and JULIAN ALPS, price 2s. 6d.

London: LONGMANS and CO.

www.ingramcontent.com/pod-product-compliance
Lightning Source LLC
Chambersburg PA
CBHW022128160426
43197CB00009B/1199